Krebs cycle

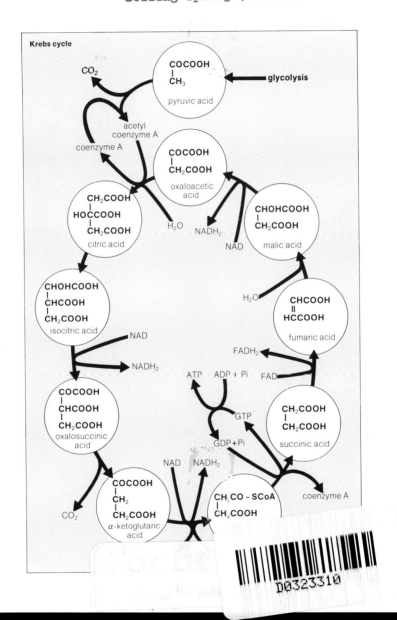

NEIL CURTIS

LONGMAN ILLUSTRATED DICTIONARY OF BIOLOGY

living organisms in all forms
explained and illustrated

LONGMAN  YORK PRESS

YORK PRESS
Immeuble Esseily, Place Riad Solh, Beirut.

LONGMAN GROUP UK LIMITED
Burnt Mill, Harlow, Essex.

First published 1985
Second impression 1988

ISBN 0 582 89255 4

Illustrations by Charlotte Kennedy and Jane Cheswright
with Philip Corke and Brian Ainsworth
Photocomposed in Britain by Prima Graphics, Camberley, Surrey, England.
Printed and bound in Lebanon by Typopress, Beirut.

Contents

How to use the dictionary

This dictionary contains over 1800 words used in the biological sciences. These are arranged in groups under the main headings listed on pp. 3-4. The entries are grouped according to the meaning of the words to help the reader to obtain a broad understanding of the subject.

At the top of each page the subject is shown in bold type and the part of the subject in lighter type. For example, on pp. 18 and 19:

18 · **THE CELL**/CARBOHYDRATES

THE CELL/CARBOHYDRATES · **19**

In the definitions the words used have been limited so far as possible to about 1500 words in common use. These words are those listed in the 'defining vocabulary' in the *New Method English Dictionary* (fifth edition) by M. West and J. G. Endicott (Longman 1976). Words closely related to these words are also used: for example, *characteristics*, defined under *character* in West's *Dictionary*.

---

1. To find the meaning of a word

Look for the word in the alphabetical index at the end of the book, then turn to the page number listed.

In the index you may find words with a letter or number at the end. These only occur where the same word appears twice in the dictionary: [a] indicates a word which is defined as it relates to animals and [p] a word defined as it relates to plants. For example, **cone**

cone[a] is part of the retina in the eye;

cone[p] is a reproductive structure in some plants.

The numbers also indicate a word which is defined twice in different contexts. For example, **translocation**

translocation[1] is the transport of materials in plants;

translocation[2] is a kind of chromosome mutation.

The description of the word may contain some words with arrows in brackets (parentheses) after them. This shows that the words with arrows are defined near by.

(↑) means that the related word appears above or on the facing page;

(↓) means that the related word appears below or on the facing page.

A word with a page number in brackets after it is defined elsewhere in the dictionary on the page indicated. Looking up the words referred to may help in understanding the meaning of the word that is being defined.

The explanation of each word usually depends on knowing the meaning of a word or words above it. For example, on p. 178 the meaning of *spore mother cell*, *microsporangium*, and the words that follow depends on the meaning of the word *spore*, which appears above them. Once the earlier words are understood those that follow become easier to understand. The illustrations have been designed to help the reader understand the definitions but the definitions are not dependent on the illustrations.

2. To find related words

Look in the index for the word you are starting from and turn to the page number shown. Because this dictionary is arranged by ideas, related words will be found in a set on that page or one near by. The illustrations will also help to show how words relate to one another.

For example, words relating to principles of classification are on pp. 40–41. On p. 40 *classification* is followed by words used to describe taxonomy and the binomial system and illustrations showing the different taxa involved in the classification of a species and the binomial system; p. 41 continues to explain and illustrate classification, explaining natural and artificial classifications and illustrating the relationships between the major groups of organisms.

3. As an aid to studying or revising

The dictionary can be used for studying or revising a topic. For example, to revise your knowledge of gas exchange, you would look up *gas exchange* in the alphabetical index. Turning to the page indicated, p. 112, you would find *respiration*, *respiratory quotient*, *breathing*, *gas exchange*, and so on; on p. 113 you would find *air*, *gill*, *gill filament*, and so on. Turning over to p. 114 you would find *counter current exchange system* etc.

In this way, by starting with one word in a topic you can revise all the words that are important to this topic.

4. To find a word to fit a required meaning

It is almost impossible to find a word to fit a meaning in most dictionaries, but it is easy with this book. For example, if you had forgotten the word for the outer whorl of the perianth of a flower, all you would have to do would be to look up *perianth* in the alphabetical index and turn to the page indicated, p. 179. There you would find the word *calyx* with a diagram to illustrate its meaning.

5. Abbreviations used in the definitions

abbr	abbreviated as	p.	page
adj	adjective	pl.	plural
e.g.	*exempli gratia* (for example)	pp.	pages
etc	*et cetera* (and so on)	sing.	singular
i.e.	*id est* (that is to say)	v	verb
n	noun	=	the same as

THE
DICTIONARY

cell theory an idea, developed in 1839, by
   Theodore Schwann, which states that all living
   organisms are made up of individual cells and
   that it is in these cells and by their division that
   processes such as growth and reproduction
   (p. 173) take place.

cell (*n*) the basic unit of a plant or animal. It is an
   individual, usually microscopic (↓) mass of
   living matter or protoplasm (p. 10). An animal
   cell consists of a nucleus (p. 13), which
   contains the chromosomes (p. 13), the
   cytoplasm (p. 10) which is usually a viscous
   fluid or gel surrounded by a very thin skin, the
   plasma membrane (p. 13). A plant cell is similar
   except that it is surrounded by a cellulose (p. 19)
   cell wall (↓) and has a fluid-filled vacuole (p. 11).

cell wall the non-living external layer of a cell in
   plants. It is comparatively rigid but slightly
   elastic and provides support for the cell. There
   may be a primary cell wall (p. 14) composed of
   cellulose (p. 19) and calcium pectate and, in
   older plants, a secondary cell wall (p. 14) made
   of layers of cellulose containing other
   substances, such as the woody lignin (p. 19).

organelle (*n*) any part of a cell, such as the
   nucleus (p. 13) or flagellum (p. 12), that has a
   particular and specialized function.

prokaryote (*n*) a cell in which the chromosomes
   (p. 13) are free in the cytoplasm (p. 10) and not
   enclosed in a membrane (p. 14): there is no
   nucleus. Bacteria (p. 42) and blue-green algae
   (p. 43) are prokaryotes.

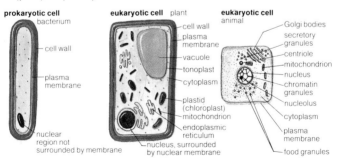

prokaryotic cell
   bacterium
   cell wall
   plasma
   membrane
   nuclear
   region not
   surrounded by membrane

eukaryotic cell   plant
   cell wall
   plasma
   membrane
   vacuole
   tonoplast
   cytoplasm
   plastid
   (chloroplast)
   mitochondrion
   endoplasmic
   reticulum
   nucleus, surrounded
   by nuclear membrane

eukaryotic cell
   animal
   Golgi bodies
   secretory
   granules
   centriole
   mitochondrion
   nucleus
   chromatin
   granules
   nucleolus
   cytoplasm
   plasma
   membrane
   food granules

optical microscope

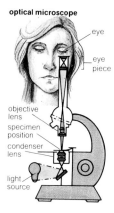

eye

eye piece

objective lens

specimen position

condenser lens

light source

electron microscope

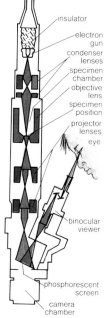

insulator

electron gun

condenser lenses

specimen chamber

objective lens

specimen position

projector lenses

eye

binocular viewer

phosphorescent screen

camera chamber

eukaryote (*n*) a cell in which the nucleus (p. 13) is separated from the cytoplasm (p. 10) by a nuclear membrane (p. 13). All organisms, except bacteria (p. 42) and blue-green algae (p. 43) are composed of eukaryotic cells.

unicellular (*adj*) of an organism consisting of one cell only.

multicellular (*adj*) of an organism composed of many cells.

cytology (*n*) the study or science of cells and their activities.

microscopy (*n*) the study, using a microscope (↓), of organisms too small to be seen with the naked eye.

microscope (*n*) an instrument used to give a magnified image of an object that is too small to be seen with the naked eye.

optical microscope a microscope (↑) in which light is passed through the object to be enlarged and passed to the eye through an objective lens system and an eyepiece. This instrument can magnify an object by a maximum of about 1500 times. For larger magnifications, an electron microscope (↓) must be used.

electron microscope a microscope (↑) which can be used to magnify objects by greater than 1500 times and to as much as 500 000 times by using electrons, which have a smaller wavelength than light, to examine the object.

ultrastructure (*n*) the structure of an object which can only be resolved using an electron microscope (↑).

sectioning (*n*) the cutting of an extremely thin slice of tissue (p. 83) which can then be examined using a microscope (↑). The tissue is first frozen or embedded in a material such as paraffin wax before it is cut. **section** (*n*).

microtome (*n*) an instrument used to cut very thin slices of a material.

staining (*n*) a method of examining particular structures inside cells by making parts of the cells opaque to light or electrons using chemicals. Certain kinds of staining materials will stain different structures e.g. iodine stains starch (p. 18).

centrifugation (*n*) a method of separating substances of different densities by accelerating them, usually in a rotating container (centrifuge), for quite long periods. Cells may be broken open and suspended in a liquid before centrifugation so that, after centrifugation, the solid particles, the sediment, will fall to the bottom of the container while a supernatant fluid will be left behind above the sediment

dialysis (*n*) a method of separating small molecules from larger molecules in a mixed solution (p. 118) by separating the solution from water by a membrane (p. 14) through which the small molecules will diffuse (p. 119) leaving behind the larger molecules which are too big to pass through the membrane.

chromatography (*n*) a method of separating mixtures of substances, such as amino acids (p. 21), by making a solution (p. 118) of the substances and allowing the substances to be absorbed (p. 81) and flow through a medium such as paper. The different substances will travel at different rates and so be separated.

chromatogram (*n*) the column or strip of solid on which substances have been separated by chromatography (↑).

electrophoresis (*n*) a method of separating mixtures of substances by suspending them in water and subjecting them to an electrical charge. Different substances will move in different directions and at different rates in response to the charge.

protoplasm (*n*) the contents of a cell.

cytoplasm (*n*) all the protoplasm (↑), or material, inside a cell other than the nucleus (p. 13), which can be thought of as alive. It is usually a viscous fluid or gel containing other organelles (p. 8), such as the Golgi body (↓). **cytoplasmic** (*adj*).

ribosome (*n*) a particle of protein (p. 21) and RNA (p. 24) which is contained in the cytoplasm (↑). Under the control of the DNA (p. 24) in the nucleus (p. 13), protein is produced on the ribosomes by linking together amino acids (p. 21). Ribosomes often occur in groups or chains.

dialysis

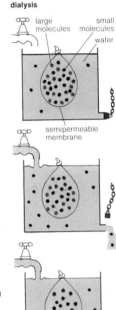

large molecules
small molecules
water
semipermeable membrane

paper chromatography

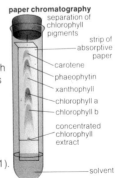

separation of chlorophyll pigments
strip of absorptive paper
carotene
phaeophytin
xanthophyll
chlorophyll a
chlorophyll b
concentrated chlorophyll extract
solvent

endoplasmic reticulum

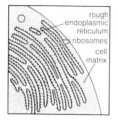

smooth endoplasmic reticulum

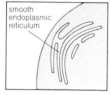

mitochondrion

cristae   matrix

smooth outer membrane

vacuole

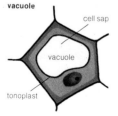

cell sap

vacuole

tonoplast

endoplasmic reticulum a meshwork of parallel, interconnected cavities within the matrix (p. 88) of a cell. These are bounded by unit membranes (p. 14) which are continuous with the nuclear membrane (p. 13). **ER** (*abbr*).

ER = endoplasmic reticulum (↑).

rough ER ER (↑) which is covered on the cytoplasmic (↑) side with ribosomes (↑).

smooth ER ER (↑) with no ribosomes (↑).

Golgi body a group or groups of flattened cavities within the cytoplasm (↑) of a cell bounded by membranes (p. 14) and connected with the ER (↑). It is similar to smooth ER (↑) but may be used for linking carbohydrates (p. 17) to proteins (p. 21), and it is associated with secretion (p. 106).

mitochondria (*n.pl.*) rod-shaped bodies in the cytoplasm (↑) of a cell. They are bounded by two unit membranes (p. 14) of which the inner is folded inwards into crests or cristae. Cell respiration (p. 30) and energy production take place in these bodies and there are more of them in cells that use a lot of energy.

lysosomes (*n.pl.*) spherical bodies that occur in the cytoplasm (↑) of cells. They are bounded by membranes (p. 14) and contain enzymes (p. 28) which may be released to destroy unwanted organelles (p. 8) or even whole cells.

microtubule (*n*) a fibrous (p. 143) structure made of protein (p. 21) found in the cytoplasm (↑). They may occur singly or in bundles. Their function may be cellular transport, e.g. the spindle (p. 37) fibres in nuclear division (p. 35).

microfilament (*n*) a very fine, thread-like structure made of protein (p. 21) which occurs in the cytoplasm (↑) of most cells.

fibril (*n*) a small fibre (p. 143) or thread-like structure.

vacuole (*n*) a droplet of fluid bounded by a membrane (p. 14) or tonoplast (↓) and contained within the cells of plants and animals except bacteria (p. 42) and blue-green algae (p. 43).

tonoplast (*n*) the inner plasma membrane (p. 13) of a cell in plants which separates the vacuole (↑) from the cytoplasm (↑).

protoplast (*n*) the protoplasmic (↑) material between the tonoplast (↓) and the plasma membrane (p. 13).

cell sap fluid contained in a plant vacuole (p. 11).
plastid (*n*) in plants, except bacteria (p. 42), blue-green algae (p. 43), and fungi (p. 46), a membrane (p. 14) bounded body in the cytoplasm (p. 10) which contains DNA (p. 24), pigments (p. 126), and food reserves.
chloroplast (*n*) in plants only, the plastid (↑) containing chlorophyll (↓) and the site of photosynthesis (p. 93). It is always bounded by a double unit membrane (p. 14).

chloroplast

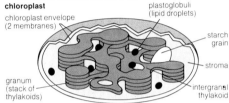

chloroplast envelope (2 membranes)
plastoglobuli (lipid droplets)
starch grain
stroma
granum (stack of thylakoids)
intergranal thylakoid

chlorophyll (*n*) a green pigment (p. 126) found in the chloroplasts (↑) of plants which is important in photosynthesis (p. 93). There are two forms of chlorophyll; chlorophyll *a* and chlorophyll *b*.
leucoplast (*n*) a colourless plastid (↑) e.g. starch (p. 18) grains.
stroma (*n*) the matrix (p. 88) within a chloroplast (↑) containing starch (p. 18) grains and enzymes (p. 28).
grana (*n.pl.*) disc-shaped, flattened vesicles (↓) in the stroma (↑) of a chloroplast (↑) holding the chorophyll (↑). **granum** (*sing.*).
vesicle (*n*) a thin-walled drop-like structure or a cavity containing fluid.
lamella (*n*) a thin plate-like structure. **lamellae** (*pl.*).
cilium (*n*) in animals and a few plants, a fine thread which projects from the surface of a cell and moves the fluid surrounding it by a beating or rowing action. **cilia** (*pl.*).
flagellum (*n*) a fine, long thread which projects from the surface of a cell and moves with an undulating action. In bacteria, the flagellum provides locomotion (p. 143) by a whip-like action during part of their life history. It is longer than a cilium (↑). **flagella** (*pl.*).

cilium

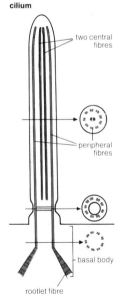

two central fibres
peripheral fibres
basal body
rootlet fibre

basal body a tiny, rod-shaped body situtated at the base of a cilium (↑) or flagellum (↑) composed of nine fibrils (p. 11) arranged in a ring at the edge of the cilium. There are also two central fibrils which do not form part of the basal body.

microvilli (*n.pl.*) finger-like projections from the surface of the plasma membrane (↓) of a cell which improve the absorption (p. 81) powers of the cell by increasing its surface area. *See also* villi (p. 103).

nucleus

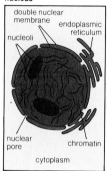

double nuclear membrane

endoplasmic reticulum

nucleoli

nuclear pore

chromatin

cytoplasm

nucleus (*n*) a body present within the cells of eukaryotic (p. 9) organisms which contains the chromosomes (↓) of the organism. **nuclear** (*adj*).

nuclear membrane the firm, double unit membrane (p. 14) surrounding the nucleus (↑) and separating it from the cytoplasm (p. 10) while allowing the exchange of materials between the nucleus and the cytoplasm through its pores (p. 120).

nucleolus (*n*) a small, round dense body, one or two of which may be present within the nucleus (↑). It is rich in RNA (p. 24) and protein (p. 21) but is not contained within a membrane (p. 14).

chromosome (*n*) a rod- or thread-shaped body occurring within the nucleus (↑) and which is readily stained by various dyes, hence its name. A chromosome is composed of DNA (p. 24) or RNA (p. 24), and protein (p. 21). Each chromosome is in the form of a long helix (p. 25) of DNA. Chromosomes mostly occur in pairs called homologous chromosomes (p. 39). They are composed of thousands of genes (p. 196) which give rise to and control particular characteristics and functions of the organism, such as eye colour, and which are passed on through the offspring by inheritance (p. 196). Each organism has a consistent number of chromosomes, e.g. in human cells, 23 pairs.

chromatin (*n*) a granular compound of nucleic acid (p. 22) and protein (p. 21) in the chromosome (↑) and which is strongly stained by certain dyes.

plasma membrane an extremely thin membrane (p. 14) separating the cell from its surroundings. It allows the transfer of substances between the cell and its surroundings.

plasmalemma (*n*) the plasma membrane (p. 13) or cell membrane (↓).

unit membrane the common structure, divided into three layers, of the plasma membrane (p. 13), or other membranes, such as the endoplasmic reticulum (p. 11). It comprises a monomolecular film (↓) and a bimolecular leaflet (↓).

monomolecular film a layer, one molecule thick, of protein (p. 21) which occurs either side of the bimolecular leaflet (↓) and forms part of the organization of the unit membrane (↑). Stained and under an electron microscope (p. 9), it appears as a dark stratum (↓).

bimolecular leaflet a layer, two molecules thick, of lipid (p. 20) which is found between two monomolecular films (↑) and forms part of the organization of the unit membrane (↑). Stained and under an electron microscope (p. 9) it appears as a light stratum (↓).

stratum (*n*) a layer. **strata** (*pl.*).

phagocytosis (*n*) the process in which a cell flows around particles in its surroundings and takes them into the cytoplasm (p. 10) to form a vacuole (p. 11).

pinocytosis (*n*) a process in which a cell folds back within itself and surrounds a tiny drop of fluid in its surroundings and takes it into the cytoplasm (p. 10) to form a vesicle (p. 12).

middle lamella in plants, the material which is laid down between adjacent cell walls (p. 8) and sticks the cells together. It is laid down as new cells form.

primary cell wall the first cell wall (p. 8) of a young cell which is laid down as the new cell forms. *See also* secondary cell wall (↓).

secondary cell wall a cell wall (p. 8) which is laid down inside the primary cell wall (↑). It surrounds some of the cells in older plants.

pit (*n*) a small area of the secondary cell wall (↑) which has remained almost unthickened or absent during the formation of the secondary wall. It allows substances to pass between the cells. The pits in one cell correspond in position with the pits in a neighbouring cell.

unit membrane

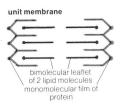

bimolecular leaflet
of 2 lipid molecules
monomolecular film of
protein

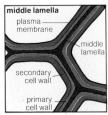

middle lamella

plasma
membrane

middle
lamella

secondary
cell wall

primary
cell wall

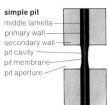

simple pit
middle lamella
primary wall
secondary wall
pit cavity
pit membrane
pit aperture

bordered pit

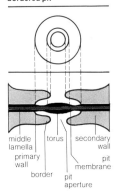

middle
lamella

torus

secondary
wall

primary
wall

pit
membrane

border

pit
aperture

plasmodesmata
plasmodesma comprising
cytoplasm and tube of
endoplasmic reticulum

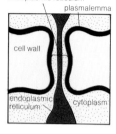

hydrogen bond between
water molecules

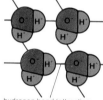

hydrogen bond (attraction
between positive hydrogen
atom and negative oxygen
atom)

plasmodesmata (*n.pl.*) fine threads of cytoplasm (p. 10) which connect the cytoplasm of neighbouring cells, and may be grouped through the membranes (↑) of pits (↑). Plasmodesmata run through narrow pores (p. 120) in the cellulose (p. 19) cell wall (p. 8). **plasmodesma** (*sing.*).

biochemistry (*n*) the study or science of the chemical substances and their reactions in animals and plants.

organic compound any substance which is a compound of carbon, except for the oxides and carbonates of carbon, and from which all living things are made. Oxygen and carbon are the main components of organic compounds.

inorganic compound a compound which, except for the oxides and carbonates, does not contain carbon, and which is not an organic compound (↑). Salt is an example of an inorganic compound.

hydrogen bond a bond which holds one molecule of water to another molecule making water more stable than it otherwise would be. A molecule of water consists of two hydrogen atoms bonded to one oxygen atom by sharing electrons. The resulting molecule is weakly polar with hydrogen atoms being positively charged and the oxygen negatively charged. These polar molecules are weakly attracted to one another.

acid (*n*) a substance that releases hydrogen (H^+) ions in a watery solution (p. 118) or accepts electrons in chemical reactions. An acid can be an inorganic compound (↑), such as hydrochloric acid, HCl or an organic compound (↑) such as ethanoic acid, CH_3COOH. The acidity of a solution can be measured on the pH scale (-logH$^+$ concentration). **acidic** (*adj*).

base[1] (*n*) a substance that releases hydroxyl (OH$^-$) ions in a watery solution (p. 118) or gives up electrons in chemical reactions, e.g. sodium hydroxide, NaOH. **basic** (*adj*).

pH *see* acid (↑).

buffer (*n*) a substance which helps a solution (p. 118) to resist a change in pH (↑) when an acid (↑) or base (↑) is added to the solution. Many biological fluids function as buffers.

condensation example of a condensation reaction

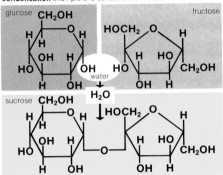

condensation (*n*) a reaction whereby two simple organic compounds (p. 15), such as glucose (↓) and fructose (↓), combine to form another compound, such as sucrose (p. 18) and a molecule of water.

hydrolysis (*n*) a reaction in which water combines with an organic compound (p. 15), such as sucrose (p. 18), to form two new organic compounds, such as glucose (↓) and fructose (↓). The reverse of condensation (↑).

molecular biology the study or science of the structure and activities of the molecules which make up animals and plants.

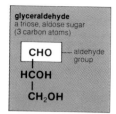

glyceraldehyde
a triose, aldose sugar
(3 carbon atoms)

CHO — aldehyde group
HCOH
CH_2OH

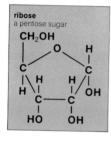

ribose
a pentose sugar

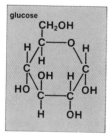

glucose

fructose

carbohydrate (*n*) an organic compound (p. 15) containing the elements carbon, hydrogen, and oxygen with the general formula $(CH_2O)_n$. Carbohydrates are essential in the metabolism (p. 26) of all living things.

monosaccharide (*n*) a carbohydrate (↑) composed of small molecules. Monosaccharides are the building blocks from which disaccharides (p. 18) and polysaccharides (p. 18) are built. Common monosaccharides found in cells contain from three to seven carbon atoms. A monosaccharide is the simplest sugar and, if further broken down, ceases to be a sugar.

sugar (*n*) the simplest carbohydrate (↑), a mono-, di- or polysaccharide (p. 18).

triose sugar a monosaccharide (↑) in which n for the general formula of the carbohydrate (↑) is 3. Glyceraldehyde is a triose sugar with the formula $C_3H_6O_3$.

pentose sugar a monosaccharide (↑) in which n for the general formula of the carbohydrate (↑) is 5. Ribose (p. 22) is a pentose sugar with the formula $C_5H_{10}O_5$.

hexose sugar a monosaccharide (↑) in which n for the general formula of the carbohydrate (↑) is 6. Glucose (↓) is a hexose sugar with the formula $C_6H_{12}O_6$. The atoms of hexose sugars may be arranged differently to give different types of sugars e.g. glucose and fructose (↓).

glucose (*n*) a hexose sugar (↑) which is widely found in animals and plants. Glucose provides a major source of energy in living things by being oxidized (p. 32) during respiration (p. 112) into carbon dioxide and water, releasing energy. In plants it is the product of photosynthesis (p. 93) and is stored as starch (p. 18) while in animals it is produced by the digestion (p. 98) of disaccharides (p. 18) and polysaccharides (p. 18) and is stored as glycogen (p. 19). Glucose combines with fructose (↓) to form sucrose (p. 18) by condensation (↑).

fructose (*n*) a hexose sugar (↑) which is widely found in plants. It combines with glucose (↑) to form sucrose (p. 18) by condensation (↑).

galactose (*n*) a hexose sugar (p. 17) which is a
   constituent of lactose (↓) and is found in many
   plant polysaccharides (↓) as well as in animal
   protein (p. 21)-polysaccharide combinations.
disaccharide (*n*) a carbohydrate (p. 17) which
   results from the combination of two
   monosaccharides (p. 17) by condensation
   (p. 16), e.g. maltose (↓) and sucrose (↓).

glucose + fructose

maltose (*n*) a disaccharide (↑) which is formed
   from the condensation (p. 16) of two molecules
   of glucose (p. 17). It is a product of the breakdown
   of starch (↓) during germination (p. 168) in
   plants, and digestion (p. 98) in animals. Also
   known as **malt sugar.**
sucrose (*n*) a disaccharide (↑) which is a
   compound of one molecule of glucose (p. 17)
   and one molecule of fructose (p. 17). It is
   widespread in plants but not in animals. Also
   known as **cane sugar**.
lactose (*n*) a disaccharide (↑) which is a
   compound of one molecule of glucose (p. 17)
   and one molecule of galactose (↑). It occurs
   in the milk of mammals (p. 80). Also known as
   milk sugar.
polysaccharide (*n*) a carbohydrate (p. 17) which
   results from the combination of more than two
   monosaccharides (p. 17) by condensation
   (p. 16). A polysaccharide has the general
   formula $(C_6H_{10}O_5)_n$.
starch (*n*) a polysaccharide (↑) which forms one
   of the main food reserves of green plants. It is
   found in the leucoplasts (p. 12). It stains blue-
   black with iodine.

polysaccharide
e.g. starch (amylopectin)

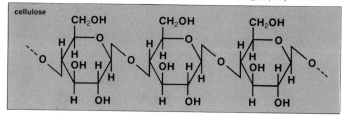

cellulose microfibrils in
surface view of plant cell
wall (×24,000)

glycogen (*n*) a polysaccharide (↑) stored by
animals and by fungi (p. 46). It is made up of
many glucose (p. 17) molecules. In vertebrates
(p. 74) it is present in large quantities in the liver
(p. 103) and muscles (p. 143).

cellulose (*n*) a long-chain polysaccharide (↑)
made up of units of glucose (p. 17). It is used
for structural support and is the main
component of the cell wall (p. 8) in plants.

cellulose

lignin (*n*) a complex organic compound (p. 15)
whose structure is not fully understood. With
cellulose (↑) it forms the chief components of
wood in trees. It is laid down in the cell walls
(p. 8) of sclerenchyma (p. 84), xylem (p. 84)
vessels, and tracheids (p. 84). It stains red with
acidified phloroglucinol. **lignified** (*adj*).

lipid (*n*) any of a number of organic compounds (p. 15) found in plants and animals with very different structures but which are all insoluble in water and soluble in substances like ethoxyethane (ether) and trichloromethane (chloroform). It is formed by the condensation (p. 16) of glycerol (↓) and fatty acids (↓). Lipids have a variety of functions including storage, protection, insulation, waterproofing, and as a source of energy.

fat (*n*) a lipid (↑) formed from the alcohol glycerol (↓) and one or more fatty acids (↓). It is solid at room temperature.

oil (*n*) a lipid (↑) formed from the alcohol glycerol (↓) and one or more fatty acids (↓). It is liquid at room temperature.

glycerol (*n*) an alcohol with the formula $C_3H_8O_3$ which is formed by the hydrolysis (p. 16) of a fat. It is a sweet, sticky, odourless, colourless liquid. Its modern name is propane-1,2,3,-triol.

fatty acid an organic acid (p. 15) with the general formula ($R(CH_2)_nCOOH$) which can be united with glycerol (↑) by condensation (p. 16) to give a lipid (↑). In living organisms, fatty acids usually have unbranched chains and an even number of carbon atoms.

triglyceride (*n*) the major component of animal and plant lipids (↑). It is derived from glycerol (↑) which has three reactive hydroxyl groups, by condensation (p. 16) with three fatty acids (↑).

phospholipid (*n*) a lipid (↑) which contains a phosphate group as an essential part of the molecule. It is derived from glycerol (↑) attached to two fatty acids (↑), a phosphate group, and a nitrogenous base. Phospholipids are essential components of cell membranes (p. 14).

saturated (*adj*) of a carbon chain, such as that in a fatty acid (↑), in which each carbon atom is attached by single bonds to carbons, hydrogen atoms, or other groups. It is unreactive.

unsaturated (*adj*) of a carbon chain, such as that in a fatty acid (↑), in which carbon atoms are attached to other groups with at least one double or triple bond. An unsaturated fatty acid is reactive and may be essential to maintain a vital structure or function in an organism.

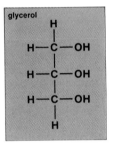

glycerol

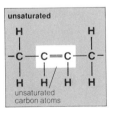

unsaturated

unsaturated carbon atoms

primary, secondary, tertiary and quaternary structure of proteins

primary structure   secondary structure

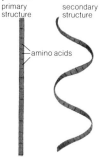

amino acids

tertiary structure

quaternary structure

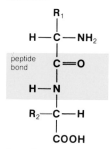

peptide bond between amino acids

R₁ and R₂ are side groups

steroid (*n*) a complex, saturated (↑) hydrocarbon in which the carbon atoms are arranged in a system of rings. All steroids are chemically similar but may have very different functions in organisms. The most common steroid in animals is cholesterol.

protein (*n*) a very complex organic compound (p. 15) made up of large numbers of amino acids (↓). Proteins make up a large part of the dry weight of all living organisms.

amino acid an organic compound (p. 15) with an amino group of atoms (–NH₂) and acidic (p. 15) carboxyl (–COOH) groups of atoms on the molecule. The general formula is RCHNH₂COOH with R representing a hydrogen or carbon chain. There are more than twenty naturally occurring amino acids with different R groups. Hundreds of thousands of amino acids are linked together to form a protein (↑). *See also* the diagram, amino acids and the genetic code on p. 204.

dipeptide (*n*) an organic compound (p. 15) which results from linking together two amino acids (↑) by condensation (p. 16).

polypeptide (*n*) an organic compound (p. 15) which results from linking together many amino acids (↑) by condensation (p. 16). In turn, polypeptides may be linked together to form proteins (↑).

peptide bond the link which joins one amino acid to the carboxyl (–COOH) group of another, resulting in the formation of a dipeptide (↑) or polypeptide (↑). A peptide bond can only be broken by the action of a hot acid (p. 15) or alkali.

conjugated protein a protein (↑) which occurs in combination with a non-protein or prosthetic group (p. 30). Haemoglobin (p. 126) is an example of a conjugated protein.

globular protein a protein (↑) which, because of the positive and negative charge on it, forms a complex three-dimensional structure as the opposite charges are attracted together and form weak bonds. A hormone (p. 130) is an example of a globular protein.

fibrous protein a protein (p. 21) which occurs
as long parallel chains with cross links. Fibrous
proteins are insoluble and are used for support
and other structural purposes. Keratin in hair,
hooves, feathers etc is an example of a fibrous
protein.

colloid (*n*) a substance, such as starch (p. 18),
that will not dissolve or be suspended in a liquid
but which is dispersed in it.

nucleic acid a large, long-chain molecule
composed of chains of nucleotides (↓) and
found in all living organisms. The carrier of
genetic (p. 196) information.

nucleotide (*n*) an organic compound (p. 15)
formed from ribose (↓), phosphoric acid (↓), and
a nitrogen base (↓).

ribose (*n*) a monosaccharide (p. 17) or pentose
sugar (p. 17) which forms an essential part of a
nucleotide (↑).

deoxyribose (*n*) a monosaccharide (p. 17) with
one less oxygen than ribose (↓).

phosphoric acid an inorganic compound (p. 15)
with the formula H_3PO_4 which forms an
essential part of nucleotides (↑). The phosphate
molecule from phosphoric acid forms a bridge
between two pentose (p. 17) molecules.

base[2] (*n*) a substance, such as a purine (↓) or
pyrimidine (↓), containing nitrogen, which is
attached to the main sugar-phosphate chain in
a nucleic acid (↑).

cytosine (*n*) a nitrogen base (↑) derived from
pyrimidine (↓) and found in both ribonucleic
acid (p. 24) and deoxyribonucleic acid (p. 24).

uracil (*n*) a nitrogen base (↑) derived from
pyrimidine (↓) and found only in ribonucleic
acid (p. 24).

adenine (*n*) a nitrogen base (↑) derived from
purine (↓) and found in both ribonucleic acid
(p. 24) and deoxyribonucleic acid (p. 24).

guanine (*n*) a nitrogen base (↑) derived from
purine (↓) and found in both ribonucleic acid
(p. 24) and deoxyribonucleic acid (p. 24).

thymine (*n*) a nitrogen base (↑) derived from
pyrimidine (↓) and found only in
deoxyribonucleic acid (p. 24).

nucleotide basic structure

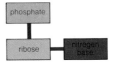

the common bases in the nucleotides of DNA and RNA

	purines	pyrimidines
DNA only		thymine
DNA and RNA	adenine guanine	cytosine
RNA only		uracil

pyrimidine (*n*) an organic compound (p. 15) with the basic formula $C_4H_4N_2$ and with a cyclic structure from which important nitrogen bases (↑) are derived.

pyrimidine base any of the several compounds related to pyrimidine (↑) and present in nucleic acids (↑).

purine (*n*) an organic compound (p. 15) with the basic formula $C_5H_4N_5$, with a double cyclic structure, from which important nitrogen bases (↑) are derived.

purine base any of several compounds related to purine (↑) and present in nucleic acids (↑).

basic molecular shape of nitrogen base

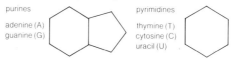

purines

adenine (A)
guanine (G)

pyrimidines

thymine (T)
cytosine (C)
uracil (U)

RNA ribonucleic acid. A nucleic acid (p. 22) consisting of a large number of nucleotides (p. 22) arranged to form a single strand. The base (p. 22) in each nucleotide is one of cytosine (p. 22), uracil (p. 22), adenine (p. 22), or guanine (p. 22). The sugar is ribose (p. 22). RNA is found in the nucleus (p. 13) of a cell and in the cytoplasm (p. 10). It usually occurs as ribosomes (p. 10) but also as *transfer RNA* and *messenger RNA*. Strands of RNA are produced in the nucleus from DNA (↓), passed to the cytoplasm, and then a ribosome is joined to the RNA. The ribosome moves along the strand of RNA and produces a polypeptide (p. 21) whose structure is controlled by the RNA. *See also* transcription and translation p. 205.

structure of portion of RNA molecule

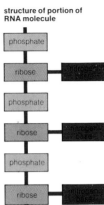

DNA deoxyxribonucleic acid. A nucleic acid (p. 22) consisting of a large number of nucleotides (p. 22) arranged to form a single strand. Usually, two strands are coiled round each other to form a double helix (↓). The base (p. 22) in each nucleotide consists of one of cytosine (p. 22), adenine (p. 22), guanine (p. 22), or thymine (p. 22). The sugar is deoxyribose (p. 22). DNA is found in the chromosomes (p. 13) of prokaryotes (p. 8) and eukaryotes (p. 9) and in the mitochondria (p. 11) of eukaryotes. It is the material of inheritance (p. 196) in almost all living organisms and is able to copy itself during nuclear divisions (p. 35).

structure of part of DNA molecule with helix unwound

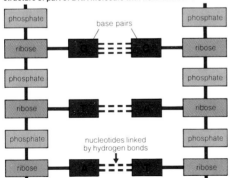

diagram of the DNA double helix

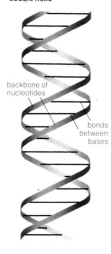

backbone of nucleotides

bonds between bases

polynucleotide chain a chain of linked nucleotides (p. 22) which makes up a nucleic acid (p. 22).

Watson-Crick hypothesis a hypothesis (p. 235) based on X-ray crystallography which suggests that DNA (↑) is a double helix (↓) of two coiled chains of alternating phosphate and sugar groups with the sugars linked by pairs of bases (p. 22).

double helix the arrangement of two helical (↓) polynucleotide chains (↑) in DNA (↑).

helix a helix is the curve that results from drawing a straight line on a plane which is then wrapped round a circular cylinder. The two helixes of DNA (↑) intertwine to form a double helix (↑) and are linked by nitrogen bases (p. 22). **helical** (adj).

base pairing the links holding the double helix (↑) of DNA (↑) together, each link consisting of a purine (p. 22) linked to a pyrimidine (p. 22) by hydrogen bonds (p. 15).

vitamin (n) the name given to a variety of organic compounds (p. 15) which are required by organisms for metabolism (p. 26) and which cannot usually be synthesized by the organism in sufficient quantities to replace that which is broken down during metabolism. See p. 238.

Benedict's test a method to determine the presence of monosaccharides (p. 17) and some disaccharides (p. 18) by adding a solution (p. 118) of copper sulphate, sodium citrate, and sodium carbonate to a solution of the sugar which produces a red precipitate (p. 26) when boiled because the sugar reduced the copper sulphate to copper (I) oxide. Sucrose (p. 18) and other non-reducing sugars do not reduce copper sulphate but it can be detected by hydrolysing (p. 16) it first into its component reducing sugars.

Fehling's test this is similar to Benedict's test (↑) but the reagent (p. 26) used is a solution (p. 118) containing copper sulphate, sodium potassium tartrate, and sodium hydroxide.

iodine test a method to determine the presence and distribution of starch (p. 18) in cells by cutting a thin section (p. 9) of the material and mounting it in iodine dissolved in potassium iodide. The starch grains turn blue-black.

emulsion test a method of testing for the presence of a lipid (p. 20) by dissolving the substance in alcohol (usually ethanol) and adding an equal volume of water. A cloudy white precipitate (↓) indicates a lipid.

alcohol/water test = emulsion test (↑).

Sudan III test a method of testing for a lipid (p. 20) which stains red with Sudan III solution.

greasemark test a method of testing for a lipid (p. 20) by taking a drop of the substance to be tested and placing it on a filter paper. When it is dry, only a lipid leaves a translucent mark when held up to the light.

translucent (*adj*) of a material that lets light pass through but through which objects cannot be seen clearly.

Millon's test a method of testing for protein (p. 21) by adding a few drops of Millon's reagent (↓) to a suspension of the protein and boiling it. The protein stains brick red.

Biuret test a method of testing for protein (p. 21) by adding an equal volume of 2 per cent sodium hydroxide solution (Biuret A) followed by 0.5 per cent copper sulphate solution (Buiret B). The protein stains purple.

emulsion (*n*) a colloidal (p. 22) suspension of one liquid in another.

suspension (*n*) a mixture in which the particles of one or more substances are distributed in a fluid.

fluid (*n*) a substance which flows, i.e. a liquid or a gas.

precipitate (*n*) an insoluble solid formed by a reaction which occurs in solution (p. 118).

reagent (*n*) a substance or solution (p. 118) used to produce a characteristic reaction in a chemical test.

metabolism (*n*) a general name for the chemical reactions which take place within the cells of all living organisms.

metabolite (*n*) any of the substances, inorganic (p. 15) or organic (p. 15) such as water or carbon dioxide, amino acids (p. 21) or vitamins (p. 25) which take part in metabolism (↑).

metabolic pathway a series of small steps in which metabolism (↑) proceeds.

metabolism chemical reations in a plant cell

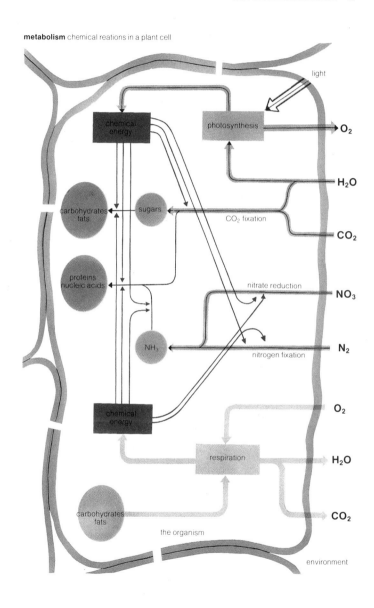

the function of enzymes in catalysis of reactions

synthesis

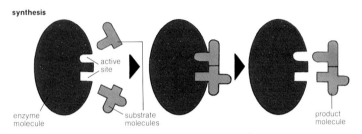

enzyme molecule · active site · substrate molecules · product molecule

enzyme (*n*) a protein (p. 21) which increases the rate at which the chemical processes of metabolism (p. 26) take place without being used up by the reaction which it affects. Enzymes are present in all living cells. They are easily destroyed by high temperatures (denatured) and require certain conditions before they will act. The rate of an enzyme-catalyzed (↓) reaction depends upon the concentration of substrate (↓) and enzyme temperature and pH (p. 15). Enzymes increase the rate of reactions by lowering the activation energy.

intracellular (*adj*) within a cell. For example, most enzyme (↑) activity is intracellular i.e. takes place within the cell that produces the enzymes.

extracellular (*adj*) outside a cell. For example, digestive (p. 98) enzymes (↑), which are extracellular in their activity may be secreted (p. 106) into the gut (p. 98) of an animal from other cells where they are produced.

in vivo 'in life' (*adj*) of all the processes which take place within the living organism itself.

in vitro 'in glass' (*adj*) of processes, such as the culture of cell tissues (p. 83) which are carried out experimentally outside the living organism, and originally derived from experiments carried out on parts of an organism in a test tube.

catalyst (*n*) any substance, such as an enzyme (↑), which increases the rate at which a chemical reaction takes place but which is not consumed by the reaction. **catalyze** (*v*).

breakdown

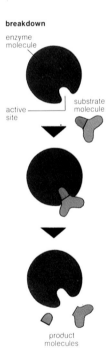

enzyme molecule · active site · substrate molecule · product molecules

enzymes control of reaction rate through substrate concentration

enzyme in excess—
reaction rapid

enzyme and
substrate
concentration
equal—reaction
rate medium

substrate in
excess—reaction
rate slow

substrate (*n*) the substance on which something
acts e.g. most enzymes (↑) only work on one
substrate each and become attached to the
substrate molecules.

active site that part of the enzyme (↑) molecule to
which specific substrate (↑) molecules become
attached.

enzyme-substrate complex the combination of
the enzyme (↑) molecule with the substrate (↑)
molecule.

lock and key hypothesis a hypothesis (p. 235)
which explains the properties of enzymes (↑) by
supposing that the particular shape of an
enzyme protein (p. 21) corresponds with the
shape of particular molecules like a lock and
key so that one enzyme will only act as a
catalyst (↑) for one specific kind of molecule.

inhibitor (*n*) a substance which slows down or
stops a reaction which is controlled by an
enzyme (↑). **inhibition** (*n*). **inhibit** (*v*).

competitive inhibition inhibition (↑) when the
substrate (↑) and the inhibitor compete for the
enzyme (↑). Also known as reversible (p. 30)
inhibition.

non-competitive inhibition inhibition (↑) when the
inhibitor combines permanently with the
enzyme (↑) so that the substrate (↑) is excluded.
Also known as non-reversible (p. 30) inhibition.

non-competitive inhibition

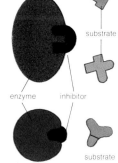

substrate

enzyme   inhibitor

substrate

reversible (*adj*) of a reaction or process that is not permanent i.e. it can work in the opposite direction. **reverse** (*v*).

non-reversible (*adj*) not reversible (↑).

cofactor (*n*) an additional inorganic compound (p. 15) which must be present in a reaction before the enzyme (p. 28) will catalyze (p. 28) it.

co-enzyme (*n*) an additional, non-protein (p. 21) organic compound (p. 15) which must be present in a reaction before the enzyme (p. 28) will catalyze (p. 28) it.

prosthetic group a non-protein (p. 21) organic compound (p. 15) which forms an essential part of the enzyme (p. 28) and which must be present in a reaction before the enzyme will catalyze (p. 28) it.

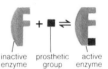

prosthetic group
role in enzyme reaction

inactive enzyme   prosthetic group   active enzyme

hydrolase (*n*) an enzyme (p. 28) which catalyzes (p. 28) hydrolysis (p. 16) reactions.

carbohydrase (*n*) an enzyme (p. 28) which catalyzes (p. 28) digestion (p. 98) reactions and aids in the breakdown of carbohydrates (p. 17).

oxidase (*n*) an enzyme (p. 28) group which catalyzes (p. 28) oxidation (p. 32) reactions.

dehydrogenase (*n*) an enzyme (p. 28) group which catalyzes (p. 28) reactions in which hydrogen atoms are removed from a sugar.

carboxylase (*n*) an enzyme (p. 28) group which catalyzes (p. 28) reactions in which carboxyl (COOH) groups are added to a substrate (p. 29).

transferase (*n*) an enzyme (p. 28) group which catalyzes (p. 28) reactions in which a group is transferred from one substrate (p. 29) to another.

isomerase (*n*) an enzyme (p. 28) group which catalyzes (p. 28) reactions in which the atoms of molecules are rearranged.

cell respiration the breakdown by oxidation (p. 32) of sugars yielding carbon dioxide, water and energy.

endergonic (*adj*) of a reaction which absorbs (p. 81) energy.

exergonic (*adj*) of a reaction which releases energy.

electron (*n*) a very small, negatively charged particle in an atom which may be raised to higher energy levels and then released during cell respiration (↑).

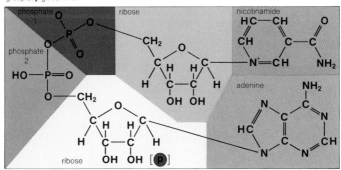

**electron carrier system
in respiration**

electron (hydrogen) carrier system a system
which operates during cell respiration (↑) in
which electrons (↑) (initially released as part of
a hydrogen atom which splits into an electron
and a proton) are collected by an electron
acceptor (↓) and passed to another electron
acceptor at lower energy levels. The energy
released in the process is used to convert ADP
(p. 33) to ATP (p. 33).

electron acceptor a molecule which functions as
a coenzyme (↑) with a dehydrogenase (↑) that
catalyzes (p. 28) the removal of hydrogen
during cell respiration (↑). It accepts electrons
(↑) and passes them on to electron acceptors
at lower energy levels.

NAD nicotinamide adenine dinucleotide. One of
the most important coenzymes (↑) or electron
acceptors (↑) concerned with cell respiration (↑).

NADP nicotinamide adenine dinucleotide
phosphate. An important coenzyme (↑) or
electron acceptor (↑) similar to NAD (↑).

**nicotinamide adenine
dinucleotide (NAD)**
adding a further phosphate
group at **p** gives NADP

oxidation (*n*) a reaction in which a substance (1) loses electrons (p. 30); (2) has oxygen added to it; or (3) has hydrogen removed from it. **oxidize** (*v*).

reduction (*n*) a reaction in which a substance (1) gains electrons (p. 30); (2) has oxygen removed from it; or (3) has hydrogen added to it. **reduce** (*v*).

cytochrome (*n*) one of a system of coenzymes (p. 30) involved in cell respiration (p. 30) having prosthetic groups (p. 30) which contain iron. Cytochromes are involved in the production of ATP (↓) by oxidative phosphorylation (p. 34).

flavoprotein (*n*) FP. An important coenzyme (p. 30) involved in cell respiration (p. 30).

vitamin B the collective name for a group of vitamins (p. 25) which play an important role in cell respiration (p. 30) by functioning as coenzymes (p. 30).

aerobic (*adj*) of a reaction, for example, respiration (p. 112) which can only take place in the presence of free, gaseous oxygen. In aerobic respiration, organic compounds (p. 15) are converted to carbon dioxide and water with the release of energy. Organisms that use aerobic respiration are called aerobes.

anaerobic (*adj*) of a reaction, for example, respiration (p. 112) which takes place in the absence of free gaseous oxygen. In anaerobic respiration organic compounds (p. 15) such as sugars are broken down into other compounds such as carbon dioxide and ethanol with a lower release of energy. Organisms that use anaerobic respiration are called anaerobes.

basal metabolism the smallest (or minimum) amount of energy needed by the body to stay alive. It varies with the age, sex and health of the organism.

BMR basal metabolic rate = basal metabolism (↑).

metabolic rate in cell respiration (p. 30) the rate at which oxygen is used up and carbon dioxide is produced.

calorific value the amount of heat produced, measured in calories, when a given amount of food is completely burned. *See also* joule (p. 97).

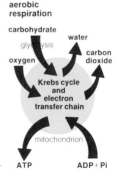

aerobic
respiration

carbohydrate

water

glycolysis

oxygen

carbon dioxide

Krebs cycle and electron transfer chain

mitochondrion

ATP

ADP + Pi

ATP adenosine triphosphate. An organic compound (p. 15) composed of adenine (p. 22), ribose (p. 22), and three inorganic phosphate groups. It is a nucleotide (p. 22) and is responsible for storing energy temporarily during cell respiration (p. 30). It is formed by the addition of a third phosphate group to ADP (↓) which stores the energy that is released when required in other metabolic (p. 26) processes.

ADP adenosine diphosphate. The organic compound (p. 15) which accepts a phosphate group to form ATP (↑).

ADP, ATP and their reactions

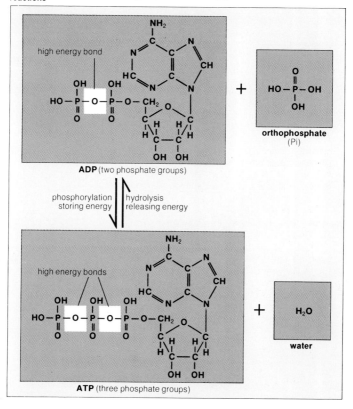

ADP (two phosphate groups)

phosphorylation storing energy | hydrolysis releasing energy

orthophosphate (Pi)

ATP (three phosphate groups)

water

phosphate bond a bond which links the phosphate groups in ATP (p. 33) and which is often misleadingly referred to as a high energy bond. Energy is stored throughout the ATP molecule but is released as the phosphate bonds are broken and other bonds are formed.

oxidative phosphorylation the process in which ATP (p. 33) is produced from ADP (p. 33) in the presence of oxygen during aerobic (p. 32) cell respiration (p. 30).

glycolysis (n) the first part of cellular respiration (p. 30) in which glucose (p. 17) is converted into pyruvic acid (↓) in the cytoplasm (p. 10) of all living organisms. It uses a complex system of enzymes (p. 28) and coenzymes (p. 30). It produces energy for short periods in the form of ATP (p. 33) when there is a shortage of oxygen. See endpaper.

pyruvic acid an organic compound (p. 15) which is formed as the end product of glycolysis (↑). For every molecule of glucose (p. 17) two molecules of pyruvic acid are formed.

Kreb's cycle a part of cellular respiration (p. 30) in which pyruvic acid (↑) in the presence of oxygen and via a complex cycle of enzyme- (p. 28) controlled reactions produces energy in the form of ATP (p. 33) and intermediates which give rise to other substances such as fatty acids (p. 20) and amino acids (p. 21). It takes place in the mitochondria (p. 11). See endpaper.

fermentation (n) a process in which pyruvic acid (↑) in the absence of oxygen uses up hydrogen atoms and so produces NAD (p. 31) allowing it to be used again in glycolysis (↑).

lactic acid fermentation fermentation (↑) from which lactic acid is produced. In higher animals this takes place especially in the muscles (p. 143) where there is an oxygen debt (p. 117).

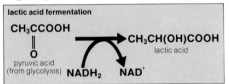

lactic acid fermentation

CH_3CCOOH ‖ O pyruvic acid (from glycolysis) → $CH_3CH(OH)COOH$ lactic acid

$NADH_2$ → NAD^+

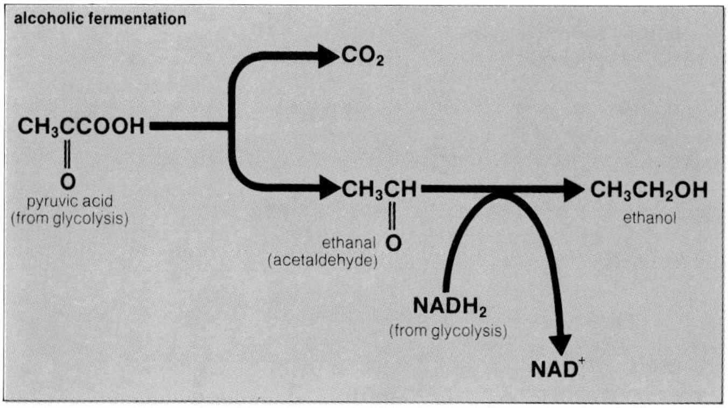

alcoholic fermentation

CH_3CCOOH

$\parallel$

O

pyruvic acid
(from glycolysis)

CO_2

CH_3CH

$\parallel$ O

ethanal
(acetaldehyde)

CH_3CH_2OH

ethanol

$NADH_2$
(from glycolysis)

NAD^+

alcoholic fermentation fermentation (↑) in which
ethanol (alcohol) and carbon dioxide are
produced. This process is made use of in the
brewing and wine-making industries in which
yeasts (p. 49) decompose sugars to provide
energy for their reproduction (p. 173) and growth.

nuclear division the process in which the nucleus
(p. 13) of a cell divides into two in the
development of new cells and new tissue
(p. 83) so that growth may occur or damaged
cells be replaced. There are two types: mitosis
(p. 37) and meiosis (p. 38).

centriole (n) a structure similar to a basal body
(p. 13). Centrioles are found outside the nuclear
membrane (p. 13) and divide at mitosis (p. 37)
forming the two ends of the spindle (p. 37).

chromatid (n) one of a pair of thread-like structures
which together appear as chromosomes (p. 13)
and which shorten and thicken during the
prophase (p. 37) of nuclear division (↑).

centromere (n) a region, somewhere along the
chromosome (p. 13), where force is exerted
during the separation of the chromatids (↑) in
mitosis (p. 37) and meiosis (p. 38).

chromomere (n) one of a number of granules of
chromatin (p. 13) which occur along a dividing
chromosome (p. 13) probably as a result of the
coiling and uncoiling within the chromatids (↑).
It appears as a 'bump' or constriction.

somatic cell any cell in a living organism other
than a germ cell (↓) and which contains the
characteristic number of chromosomes (p. 13),
normally diploid (↓), for the organism.
germ cell a cell that gives rise to a gamete
(p. 175). A cell in a living organism, other than a
somatic cell (↑), and which takes part in the
reproduction (p. 173) of the organism. It
contains only half of the characteristic number
of chromosomes (p. 13) of the organism i.e. it
is haploid (↓).

diploid and haploid stages in the life-cycle of a flowering plant

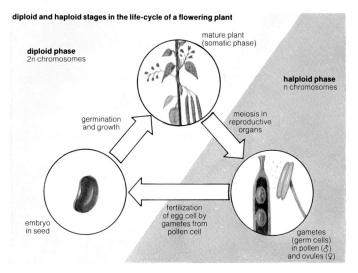

diploid phase
2n chromosomes

mature plant
(somatic phase)

halploid phase
n chromosomes

germination
and growth

meiosis in
reproductive
organs

embryo
in seed

fertilization
of egg cell by
gametes from
pollen cell

gametes
(germ cells)
in pollen (♂)
and ovules (♀)

haploid (*adj*) of a cell which has only unpaired
chromosomes (p. 13); half the diploid (↓)
number of chromosomes which are not paired
in the haploid state. Germ cells (↑) of most
animals and plants are haploid. *See also*
polyploidy etc p. 207.
diploid (*adj*) of a cell which has chromosomes
(p. 13) which occur in homologous (p. 39) pairs.
Somatic cells (↑) of most higher plants and
animals are described as diploid. Double the
haploid (↑) number.

mitosis
(only two pairs of homologous
chromosomes shown for
clarity)

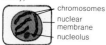

chromosomes
nuclear
membrane
nucleolus

prophase chromosomes
become visible in the nucleus,
each one duplicated into two
chromatids, joined by a
centromere

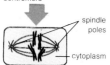

spindle
poles

cytoplasm

metaphase nuclear
membrane and nucleolus
have disintegrated. Spindle
fibres form. Chromosomes
shorter and thicker, arranged
midway between the spindle
poles.

anaphase chromatids
separate at centromeres.
Sister chromatids drawn to
opposite poles of the spindle.

telophase nuclear membrane
and nucleoli reform.
Chromosomes begin to lose
their compact structure

interphase chromosomes no
longer visible.

mitosis (*n*) the usual process of nuclear division
(p. 35) into two daughter nuclei (p. 13) during
vegetative growth. During mitosis each
chromosome (p. 13) duplicates itself, each one
of the duplicates going into separate daughter
nuclei. The daughter cells are identical to each
other and to the parent cell.

spindle (*n*) fibrous (p. 143) material which forms
from the centrioles (p. 35) during mitosis (↑)
and meiosis (p. 38). It takes part in the
distribution of chromatids (p. 35) to the daughter
cells. The chromosomes (p. 13) are arranged
at its equator (↓) during metaphase (↓).

pole (*n*) one of the two points on the spindle (↑)
which is the site of the formation of the spindle
fibres (p. 143) from the centrioles (p. 35).

equator (*n*) part of the spindle (↑) midway between
the poles (↑) to which the chromosomes (p. 13)
become attached by the spindle attachment.

interphase (*n*) a stage in the cell cycle when the
cell is preparing for nuclear division (p. 35). At
this stage, the DNA (p. 24) is replicating to
produce enough for the daughter cells.

prophase (*n*) the first main stage in nuclear
division (p. 35) in which the chromosomes
(p. 13) become visible and then the chromatids
(p. 35) appear while the nucleolus (p. 13) and
nuclear membrane (p. 13) begin to dissolve.

metaphase (*n*) a main stage in nuclear division
(p. 35) at which the nuclear membrane (p. 13)
has disappeared and the chromosomes (p. 13)
lie on the equator (↑) of the spindle (↑). Then
the chromatids (p. 35) start to move apart.

anaphase (*n*) a main stage in nuclear division
(p. 35) in which the centromeres (p. 35) divide
and the chromatids (p. 35) move to opposite
poles (↑) by the contraction of the spindle (↑).

telophase (*n*) a main stage in nuclear division
(p. 35) at which the chromatids (p. 35) arrive at
the poles (↑) and the cytoplasm (p. 10) may divide
to form two separate daughter cells in interphase
(↑). The spindle (↑) fibres (p. 143) dissolve while
the nucleolus (p. 13) and nuclear membrane
(p. 13) in each daughter cell reform and the
chromosomes (p.13) regain their thread-like form.

meiosis (*n*) nuclear division (p. 35) of a special
kind which begins in a diploid (p. 36) cell and
takes place in two stages. Each stage is similar
to mitosis (p. 37) but the chromosomes (p. 13)
are duplicated only once before the first division
so that each of the four resulting daughter cells
is haploid (p. 36). It occurs during the formation
of the gametes (p. 175).

meiosis
(cytoplasm and membrane
not shown)

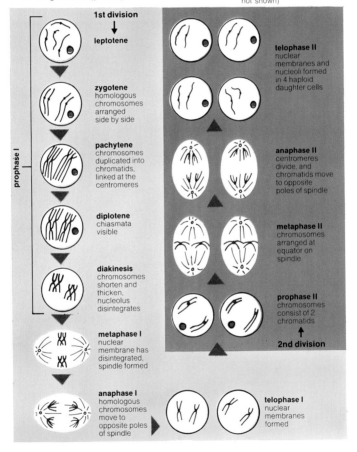

1st division

leptotene

zygotene
homologous
chromosomes
arranged
side by side

pachytene
chromosomes
duplicated into
chromatids,
linked at the
centromeres

diplotene
chiasmata
visible

diakinesis
chromosomes
shorten and
thicken,
nucleolus
disintegrates

metaphase I
nuclear
membrane has
disintegrated,
spindle formed

anaphase I
homologous
chromosomes
move to
opposite poles
of spindle

prophase I

telophase II
nuclear
membranes and
nucleoli formed
in 4 haploid
daughter cells

anaphase II
centromeres
divide, and
chromatids move
to opposite
poles of spindle

metaphase II
chromosomes
arranged at
equator on
spindle

prophase II
chromosomes
consist of 2
chromatids

2nd division

telophase I
nuclear
membranes
formed

mitosis	meiosis
occurs in somatic cells during growth and repair	occurs in the sex organs during gamete formation
no pairing or separation of homologous chromosomes	pairing and separation of homologous chromosomes
no chiasmata formed	chiasmata formed which may lead to crossing over and recombination
one separation of nuclear material i.e. separation of chromatids only	two separations of nuclear material i.e. separation of homologous chromosomes (1st division) and chromatids (2nd division)
2 daughter nuclei formed	4 daughter nuclei formed
daughter nuclei identical	daughter nuclei not identical
daughter nuclei diploid	daughter nuclei haploid

**differences between
mitosis and meiosis**

**pairs of homologous
chromosomes**

centromeres

bivalent (*n*) one of the pairs of homologous (↓) chromosomes (p. 13) which associate during the first prophase (p. 37) of meiosis (↑).

chiasmata (*n.pl.*) the points at which homologous (↓) chromosomes (p. 13) remain in contact as the chromatids (p. 35) move apart during the first prophase (p. 37) of meiosis (↑). There may be up to eight chiasmata in a bivalent (↑) pair of chromosomes. **chiasma** (*sing.*).

terminalization (*n*) the process in which the chiasmata (↑) move to the ends of the chromosomes (p. 13) during the prophase (p. 35) of meiosis (↑).

homologous chromosomes two chromosomes (p. 13) which form a pair in which the genes (p. 196) arranged along their length control identical characteristics of the organism, such as eye colour or height.

first meiotic division the first of two major stages of meiosis (↑) in which a nuclear division (p. 35) similar to mitosis (p. 37) takes place resulting in the separation of homologous (↑) chromosomes (p. 13).

second meiotic division the second of two major stages in meiosis (p. 37) in which a second nuclear division (p. 35) takes place and the two daughter cells formed from the first meiotic division (↑) each divide into two to result in four haploid (p. 36) daughter cells each containing one of the sister chromatids (p. 35).

classification (*n*) the arrangement of all living organisms into an ordered series of named and related groups. **classify** (*v*).

organisms (*n*) any living thing. Organisms can grow and reproduce (p. 175).

taxon (*n*) the general term for any group in a classification (↑) no matter what its rank (↓). **taxa** (*n.pl.*).

taxonomy (*n*) the science of classification (↑).

binomial system a system of naming every known living organism, first devised by the Swedish botanist, Carolus Linnaeus (1707–78), in which the organism is given a two-part scientific name which is usually Latinized. The first word indicates the genus (↓) while the second word indicates the species (↓). While the common names of organisms may only be understood in their place of origin, the scientific name is recognized internationally by scientists. For example, the bird with the English common name, peregrine falcon, is given the scientific name, *Falco peregrinus*.

species (*n*) a group of similar living organisms whose members can interbreed to produce fertile (p. 175) offspring but which cannot breed with other species groups. **specific** (*adj*).

genus (*n*) a group of organisms containing a number of similar species (↑). Of the scientific name, *Falco peregrinus*, *Falco* is the generic name referring to all birds that are classified (↑) as falcons.

classfication of the Peregrine falcon showing the series of ranks and their names		
rank	**scientific name of taxonomic groups (taxa)**	**common name**
kingdom	Animalia	animals
phylum	Chordata	vertebrates
class	Aves	birds
order	Falconiformes	birds of prey
family	Falconidae	falcons
genus	*Falco*	true falcons
species	*peregrinus*	Peregrine falcon

rank (*n*) one of a number of major groups into
which living organisms are classified (↑). The
largest group which contains organisms that
have different body plans from those in any
other large group is called a kingdom (↓). Each
kingdom may be further divided, on the basis of
diversity (p. 213), into a number of phyla, and
so on. The principal rank names arranged in
order from the largest groups to the most basic
are kingdom, phylum, class, order, family, genus
(↑) and species (↑).

kingdom (*n*) the highest rank (↑) or taxon (↑). Most
simply, all life can be grouped into either the
Plant or Animal kingdoms. This, however, is an
oversimplification and in this book we divide
living organisms into five kingdoms: Monera (p. 42),
Protista (p. 44), Fungi (p. 46), Plants, Animals.

artificial classification a classification (↑) in which
the organisms are arranged into groups on the
basis of apparent analogous (p. 211) similarities
which, in fact, have no common ancestry.

natural classification a classification (↑) in which
the organisms are arranged into groups on the
basis of homologous (p. 211) similarities which
demonstrate a common ancestry.

evolution and relationship
of main plant and animal
groups

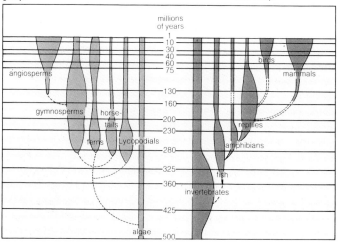

microbiology (*n*) the study or science of very small (microscopic (p. 9)) or submicroscopic living organisms. It includes bacteriology and virology.

Monera (*n*) the kingdom (p. 41) of prokaryotic (p. 8) organisms which includes the bacteria (↓) and blue-green algae (↓).

bacteria (*n.pl.*) a group of microscopic (p. 9) prokaryotic (p. 8) organisms that may be unicellular (p. 9) or multicellular (p. 9). They lack organelles (p. 8) bounded by membranes (p. 14) and contain no large vacuoles (p. 11). Most bacteria are heterotrophic (p. 92) but some are autotrophic (p. 92). Their respiration (p. 112) may be either aerobic (p. 32) or anaerobic (p. 32). Bacteria reproduce (p. 173) mainly by asexual cell division. Heterotrophic bacteria may cause disease. They are important in the decay of plant and animal tissue (p. 83) to release food materials for higher plants and in sewage breakdown. **bacterium** (*sing.*).

bacillus (*n*) a rod-shaped bacterium (↑). **bacilli** (*pl*).

coccus (*n*) a spherical-shaped bacterium (↑). **cocci** (*pl*).

streptococcus (*n*) a coccus (↑) which occurs in chains.

staphylococcus (*n*) a coccus (↑) which occurs in clusters.

spirillum (*n*) a spiral-shaped bacterium (↑). **spirilla** (*pl*).

Gram's stain a stain used in the study of bacteria (↑). Bacteria which take the violet stain are gram-positive while others that do not are gram-negative. Gram-positive bacteria are more readily killed by antibiotics (p. 233).

myxobacterium (*n*) a bacillus (↑) that has a delicate flexible cell wall (p. 8) and is able to glide along solid surfaces.

spirochaete (*n*) a spirillum (↑) which is able to move by flexing its body. Some are parasitic (p. 92) and cause diseases such as syphilis.

rickettsia (*n*) any of the various bacilli (↑) which live as parasites (p. 110) on some arthropods (p. 67) and which can be transmitted to humans causing diseases such as typhus.

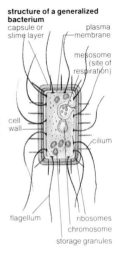

structure of a generalized bacterium

capsule or slime layer — plasma membrane — mesosome (site of respiration) — cell wall — cilium — flagellum — ribosomes — chromosome — storage granules

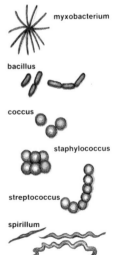

bacteria

myxobacterium

bacillus

coccus

staphylococcus

streptococcus

spirillum

filamentous blue green algae

plant viruses
spherical

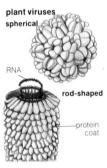

RNA

rod-shaped

protein
coat

actinomycete (n) a soil-dwelling gram-positive
(↑) bacterium (↑) with its cells arranged in
filaments (p. 181). It may be used to produce
antibiotics (p. 233) such as streptomycin.

pathogen (n) any parasitic (p. 92) bacterium (↑),
virus (↓), or fungus (p. 44) which produces
disease.

toxin (n) a poison produced by a living organism,
especially a bacterium (↑), and which may
cause the symptoms of disease which results
from the action of a pathogen (↑). **toxic** (adj).

blue-green algae a group of microscopic (p. 9)
prokaryotic (p. 8) organisms known as
Cyanophyta. They contain chlorophyll (p. 12)
and other pigments (p. 126), and are widely
distributed wherever water is present. Some
are able to take in (fix) atmospheric nitrogen
into organic compounds (p. 15).

virus (n) a pathogen (↑) which may or may not be
a living organism and which is so small that it
can only be observed with the aid of an electron
microscope (p. 9). It has no normal organelles
(p.8). A virus will only grow within its host (p.111)
but occurs as non-living chemicals outside.
Viruses are often named after their hosts to
which they are specific and the symptoms they
cause, for example, tobacco mosaic virus. A
virus consists of an outer protein (p. 21) coat
which surrounds a core of nucleic acids (p. 22).

bacteriophage (n) a virus (↑) which infects a
bacterium (↑). It consists of a head, which
contains its DNA (p. 24) or RNA (p. 24),
enclosed by a protein (p. 21) coat, and a tail
which ends in a plate that bears a number of
tail fibres (p. 143). **phage** (abbr).

life cycle of a bacteriophage

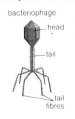

bacteriophage

head

tail

tail
fibres

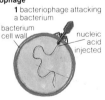

1 bacteriophage attacking
a bacterium

bacterium
cell wall

nucleic
acid
injected

2 parts of new
bacteriophages
synthesized in
bacterial cell

3 bacterium destroyed,
new bacteriophages
released

Protista (*n*) a kingdom (p. 41) of unicellular (p. 9) eukaryotic (p. 9) organisms, some of which have been previously allocated to the plant or animal kingdoms, or even both, e.g. Protozoa (↓) and unicellular algae (↓). **protistan** (*adj*).

binary fission asexual reproduction (p. 173) in which a single parent organism gives rise to two daughter organisms. The nucleus (p. 13) divides by mitosis (p. 37) followed by the division of the cytoplasm (p. 10).

algae (*n.pl.*) organisms, with a unicellular (p. 9) or simple multicellular (p. 9) body plan, that are able to manufacture their own food material by photosynthesis (p. 93). Unicellular types belong to the Protista (↑) while multicellular types, e.g. seaweeds, are regarded as plants. **alga** (*sing.*).

phycology (*n*) the science or study of algae (↑).

Protozoa (*n*) a division of the Protista (↑) in which the microscopic (p. 9) organisms are unicellular (p. 9), exist as a continuous mass of cytoplasm (p. 10), ingest (p. 98) their food and lack chloroplasts (p. 12) and cell walls (p. 8). Protozoans are widespread and important in natural communities. **protozoan** (*adj*).

Amoeba (*n*) a genus (p. 40) of Protozoa (↑). Its members consist of a single motile (p. 173) cell, able to take in food particles by engulfing them using pseudopodia (↓). **amoebae** (*pl.*).

pseudopodium (*n*) a temporary protuberance into which the cytoplasm (p. 10) of a protozoan (↑) flows and which enables it to move and feed. **pseudopodia** (*pl.*).

amoeboid movement the process of locomotion (p. 143) which results from the formation of pseudopodia (↑).

food vacuole a vacuole (p. 11) containing a food particle and a drop of water, engulfed by the pseudopodia (↑) of a protozoan (↑).

ectoplasm (*n*) the external plasma membrane (p. 13) of a protozoan (↑). A fibrous (p. 143) gel with a less granular structure than the endoplasm (↓) which it surrounds. It takes part in amoeboid movement (↑) and in cell division.

endoplasm (*n*) the cytoplasm (p. 10) of a protozoan (↑). It is more fluid and granular than ectoplasm (↑).

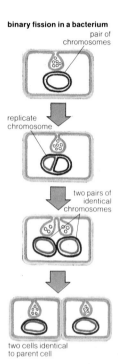

binary fission in a bacterium

pair of chromosomes

replicate chromosome

two pairs of identical chromosomes

two cells identical to parent cell

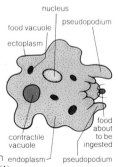

Amoeba

nucleus

food vacuole

pseudopodium

ectoplasm

contractile vacuole

food about to be ingested

endoplasm

pseudopodium

the movement of a cilium

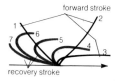

forward stroke

recovery stroke

gel (*n*) a jelly-like material

granule (*n*) a small particle **granular** (*adj*)

Paramecium (*n*) a genus (p. 40) of Protozoa (↑)
Although it is unicellular (p. 9), its organization
is more complex than that of *Amoeba*. It moves
by means of cilia (p. 12), it possesses two kinds
of nuclei (p. 13), meganuclei (↓) and micronuclei
(↓), and it reproduces (p. 173) asexually by
transverse binary fission (↑)

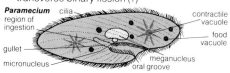

Paramecium
region of ingestion
gullet
micronucleus
cilia
contractile vacuole
food vacuole
meganucleus
oral groove

ciliate movement the process of locomotion
(p. 143) which involves the beating of stiffened
cilia (p. 12) against the water On the recovery
stroke the cilia relax so that they do not push
against the water in the reverse direction

eye spot an organelle (p. 8) which is sensitive to
light. It occurs in many Protozoa (↑)

oral groove a ciliated (p. 12) groove in *Paramecium*
(↑) into which food particles are drawn by the
beating of cilia. It leads to the gullet and the
region in which the food is ingested (p. 98)

micronucleus (*n*) the smaller of the two nuclei
(p. 13) of *Paramecium* which divides by mitosis
(p. 37) and supplies gametes (p. 175) during
conjugation (↓)

meganucleus (*n*) the larger of the two nuclei
(p. 13) of *Paramecium* (↑) which is concerned
with making protein (p. 21) for the organism

conjugation (*n*) a process of sexual reproduction
(p. 173) in *Paramecium* (↑) and other Protozoa
(↑) in which two cells temporarily come together
and exchange gametes (p. 175)

Euglena (*n*) a genus (p. 40) of Protista (↑). It
moves by means of a flagellum (p. 12) and
reproduces (p. 173) by transverse binary fission
(↑). It has no rigid cell wall (p. 8) but an elastic
transparent pellicle. It contains chloroplasts
(p. 12) by which it produces its own food
substances by photosynthesis (p. 93), but is
also able to ingest (p. 98) food through a gullet.

Euglena

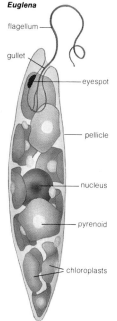

flagellum
gullet
eyespot
pellicle
nucleus
pyrenoid
chloroplasts

mycology (*n*) the science or study of fungi (↓).

Fungi (*n*) a kingdom (p. 41) of eukaryotic (p. 9) organisms that are unable to make food material by photosynthesis (p. 93). Instead they take up all their nutrients (p. 92) from their surroundings. They may be microscopic (p. 9) or quite large. They may be unicellular (p. 9) or made up of hyphae (↓). They live either as saprophytes (p. 92) or parasites (p. 110) of plants and animals. Fungi may reproduce (p. 173) sexually and asexually. **Fungus** (*sing*).

hypha (*n*) a branched, haploid (p. 36) filament (p. 181) which is the basic unit of most fungi (↑). It is a tubular structure composed of a cell wall (p. 8) with a lining of cytoplasm (p. 10) and surrounding a vacuole (p. 11). In some fungi, the hyphae (*pl*) may be divided by cross walls or septa. The cell wall is composed mainly of the material chitin (p. 49).

mycelium (*n*) a mass of hyphae (↑) which make up the bulk of a fungus (↑). **mycelia** (*pl*).

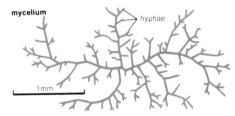

mycelium

hyphae

1 mm

coenocytic (*adj*) of hyphae (↑) which consist of tubular masses of protoplasm (p. 10) containing many nuclei (p. 13).

dikaryon (*adj*) of a hypha (↑) or mycelium (↑) made up of cells containing two haploid (p. 36) nuclei (p. 13) which divide simultaneously when a new cell is formed.

Phycomycetes (*n.pl.*) a group of Fungi (↑) which possess hyphae (↑) without septa (cross walls). Phycomycetes reproduce (p. 173) sexually by means of zygospores (↓) and asexually by means of zoospores (↓). This group includes the large genus (p. 40) of pin moulds, *Mucor*, and the related genus *Rhizopus*.

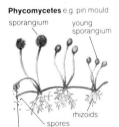

Phycomycetes e.g. pin mould

sporangium

young sporangium

rhizoids

spores

columella

zygospores
stages in formation

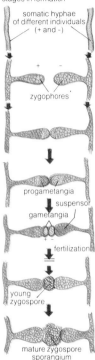

somatic hyphae
of different individuals
(+ and –)

zygophores

progametangia

suspensor
gametangia

fertilization

young
zygospore

mature zygospore
sporangium

**asci and ascospores
of Ascomycetes**

ascospores
released
explosively
from asci

asci each
with 8
ascospores

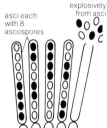

homothallic (*adj*) of the sexual reproduction
(p. 173) of certain fungi (↑) and algae (p. 44) in
which a single thallus (p. 52) produces the
opposite, differently sized gametes (p. 175) to
perform the sexual functions so that the species
is, in effect, hermaphrodite (p. 175).

heterothallic (*adj*) of the sexual reproduction
(p. 173) of certain fungi (↑) and algae (p. 44) in
which reproduction can only take place
between two genetically different thalli (p. 52)
which cannot reproduce independently. In
some fungi, the two thalli may be different in
form so that they are either male or female
while in others there may be no difference in
form but the gametes (p. 175) are different in
size between the two genetically different
strains of the same species (p. 40).

zygospore (*n*) a thick-walled resting spore
(p. 178) produced by a phycomycete (↑) during
sexual reproduction (p. 173) by the fusion of
two gametes (p. 175) called gametangia.

zoospore (*n*) the naked, flagellate (p. 12) spore
(p. 178) produced in a sporangium (p. 178)
during asexual reproduction (p. 173).

Ascomycetes (*n.pl.*) a group of Fungi (↑) which
possess hyphae (↑) with septa. Ascomycetes
reproduce (p. 173) sexually by means of
ascospores (↓) and asexually by means of
conidia (↓). *Penicillium* is an important genus
(p. 40) of Ascomycetes from which antibiotics
(p. 233) are manufactured.

ascus (*n*) a near-cylindrical or spherical cell in
which ascospores (↓) are formed. A number of
asci (*pl.*) may be grouped together into a fruit
body which is visible to the naked eye.

ascospore (*n*) the spore (p. 178) which forms in
the ascus (↑) as a result of the fusion of haploid
(p. 36) nuclei (p. 13) followed by meiosis (p. 38)
to restore the haploid state. Normally, each
ascus contains eight ascospores.

septum (*n*) a wall across a hypha (↑). **septa** (*pl.*).

conidium (*n*) a spore (p. 178) or bud which is
produced during asexual reproduction (p. 173)
from the tips of particular hyphae (↑). **conidia**
(*pl.*). See diagram on p. 48.

Basidiomycetes (*n.pl.*) a group of Fungi (p.46)
which possess hyphae (p.46) with septa. Hyphae
are often massed into substantial fruit bodies
such as mushrooms (↓) or toadstools (↓). They
reproduce (p. 173) sexually by basidiospores
(↓). *Agaricus*, including the field mushroom, is a
genus (p.40) of this group.

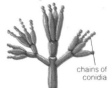
conidia e.g. *Penicillium*

chains of
conidia

mushrooms
and toadstools

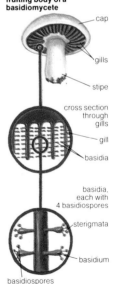
fruiting body of a
basidiomycete

cap

gills

stipe

cross section
through
gills

gill

basidia

basidia,
each with
4 basidiospores

sterigmata

basidium

basidiospores

mushroom (*n*) the common name for the fruit
body of Basidiomycetes (↑) belonging to the
order Agaricales. The name is usually used for
those species (p. 40) that are good to eat.

toadstool (*n*) the common name for the fruit body
of Basidiomycetes (↑) belonging to the order
Agaricales and which are not referred to as
mushrooms (↑). It is not necessarily a poisonous
species (p. 46).

basidiospore (*n*) a haploid (p. 36) spore (p. 178)
produced following sexual reproduction (p. 173)
and meiosis (p. 38) and borne externally on the
fruit bodies of Basidiomycetes (↑).

basidium (*n*) a club-shaped or cylindrical cell on
which the basidiospores (↑) are borne on short
stalks, usually four at a time. **basidial** (*adj*).

sterigmata (*n.pl.*) the stalks on a basidium (↑) on
which the basidiospores (↑) are borne. Each
basidial cell usually bears four sterigmata.

cap (*n*) the umbrella-shaped structure which
crowns the central stem of the larger fungi
(p. 46) forming the fruit body and in which the
spores (p. 178) are produced.

pileus (*n*) = cap (↑).

gills (*n.pl.*) the fin-like structures which occur on the underside of the cap (↑) of the fruit body of a fungus (p. 46). They bear the spore- (p. 178) producing cells or basidia (↑).

yeast (*n*) unicellular (p. 9) fungi (p. 46) which are very important in brewing and baking and as sources of proteins (p. 21) and minerals. Most yeasts are Ascomycetes (p. 47).

Fungi Imperfecti a loose grouping of fungi (p. 46) which only reproduce (p. 173) asexually.

rust (*n*) a parasitic (p. 92) basidiomycete (↑) fungus (p. 46). Rusts are serious pests of crops and may cause huge losses.

blight (*n*) a disease of plants, such as potatoes, which results from the rapid spread of the hyphae (p. 46) of fungi (p. 46), such as *Phytophthora*, through the leaves of the host (p. 110).

chitin (*n*) a horny material found in the cell walls (p. 8) of many fungi (p. 46) and composed of polysaccharides (p. 18). It is similar to the material that protects the bodies of insects (p. 69).

mycorrhiza (*n*) the symbiotic (p. 228) association (p. 227) which may occur between a fungus (p. 46) and the roots of certain higher plants, especially trees.

lichen (*n*) a symbiotic (p. 228) association (p. 227) of an alga (p. 44) and a fungus (p. 46) to form a slow-growing plant which colonizes (p. 221) such inhospitable environments (p. 218) as rocks in mountainous areas or the trunks of trees.

slime mould widely distributed fungi (p. 46) consisting of masses of protoplasm (p. 10) containing many nuclei (p. 13) and occurring in damp conditions. They reproduce by means of spores (p. 178) and are often classified (p. 40) with fungi. During part of their life history, slime moulds are able to undertake amoeboid movement (p. 44).

cellular slime mould

many cells living in soil → cells attracted to each other and flow together → cells moving through soil as a unit → spore-bearing body formed

botany (*n*) the study or science of plants or plant life.

Plantae (*n*) one of the five kingdoms (p. 41) of living organisms containing all plants that are capable of making their own food by photosynthesis (p. 93). It includes the multicellular (p. 9) algae (p. 44), Musci (p. 52), Filicales (p. 56), gymnosperms (p. 57) and angiosperms (p. 57).

Thallophyta (*n*) in the two-kingdom (p. 41) classification (p. 40) of living organisms, a division made up of all those non-animal organisms in which the body is not differentiated into stem, roots, and leaves etc. Reproduction (p. 173) takes place sexually by fusion of gametes (p. 175) and asexually by spores (p. 178). It includes bacteria (p. 42), blue-green algae (p. 43), fungi (p. 46) and lichens (p. 49).

vascular plant a plant which possesses a vascular system (p. 127) to transport water and food materials through the plant and which also provides support for the plant.

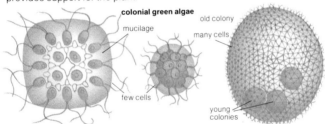

colonial green algae
mucilage
few cells
old colony
many cells
young colonies

Chlorophyta (*n*) a division of mainly multicellular (p. 9) algae (p. 44) in which the plants are mainly freshwater although there are some marine forms. These are the green algae and contain chlorophyll (p. 12) for photosynthesis (p. 93). They store food as starch (p. 18) and fats.

Chlamydomonas (*n*) a unicellular (p. 9) genus (p.40) of Chlorophyta (↑) which is found widely in freshwater ponds. They possess two flagella (p. 12) and a cup-shaped chloroplast (p. 12) contained within the cell wall (p. 8).

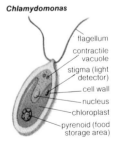

Chlamydomonas

flagellum
contractile vacuole
stigma (light detector)
cell wall
nucleus
chloroplast
pyrenoid (food storage area)

Spirogyra

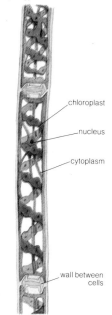

chloroplast

nucleus

cytoplasm

wall between cells

Spirogyra (*n*) a genus (p. 40) of typical filamentous (p. 181) algae of the Chlorophyta (↑). They are found in freshwater and consist of a simple chain of identical cells each containing the characteristic spiral chloroplast (p. 12).

Phaeophyta (*n*) a division of the algae (p. 44) in which the plants are largely marine, such as the large seaweeds of the shore. They contain chlorophyll (p. 12) and the brown pigment (p. 126), fucoxanthin, and they are referred to as the brown algae. They are multicellular (p. 9) and store food as sugars.

Fucus (*n*) a typical genus (p. 40) of the Phaeophyta (↑). They are the common seaweeds of the intertidal zone. Each plant is differentiated into a holdfast for clinging to the rocks, a tough stalk called a stipe, and flat fronds. They are commonly described as wracks.

bladder[D] (*n*) an air-filled sac which occurs in some members of the Phaeophyta (↑). Plants containing these bladders are commonly called bladderwracks.

conceptacle (*n*) one of the many cavities which occur at the tips of the fronds of some members of the Phaeophyta (↑) such as bladderwracks. They open by a pore (p. 120) called an ostiole, and, as well as sex organs, contain masses of sterile hairs called paraphyses.

Phaeophyta brown algae

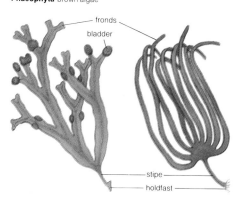

fronds

bladder

stipe

holdfast

Bryophyta (*n.pl.*) a division of the Plantae (p. 50)
including the Hepaticae (↓) Anthocerotae (↓)
and the Musci (↓) Bryophytes lack vascular
tissues (p. 83) although the stems of some
mosses have a central strand of conducting
tissue. They are mostly plants of warm places
but some are aquatic others live in desert
habitats (p. 217) or cold places and may
sometimes be the dominant form of plant life
All bryophytes show a clear alternation of
generations (p. 176) with a conspicuous, food-
independent gametophyte (p. 177) generation
and a short-lived sporophyte (p. 177) generation
dependent on the gametophyte. They are small
flattened plants with leaves and a stem but no
roots and are attached by a rhizoid (↓)

liverwort sporophyte

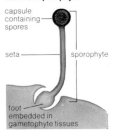

capsule
containing
spores

seta — sporophyte

foot
embedded in
gametophyte tissues

Hepaticae (*n.pl.*) liverworts A family of the
Bryophyta (↑). These are the simplest
bryophytes and may be either a flattened,
leafless gametophyte (p. 177), a thallose (↓)
liverwort, or a creeping, leafy gametophyte
known as a leafy liverwort. A typical thallose
liverwort is ribbon-like and has Y-shaped
branches They are usually aquatic, living in
damp soil or as epiphytes (p. 228).

thalloid liverwort

sporophyte
(with capsule)
thallus
(gametophyte)

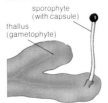

Anthocerotae (*n.pl.*) hornworts. A family of the
Bryophyta (↑). The plant consists of a lobed,
green thallus (↓) anchored by a rhizoid (↓) to the
substrate which is usually moist soil or mud·

rhizoids

thalloid
liverwort

Musci (*n.pl.*) mosses. A family of the Bryophyta
(↑). These are the most advanced bryophytes
and they either grow in erect, virtually
unbranched cushions or in feathery, creeping,
branched mats They are widely distributed
throughout the world living in damp conditions,
such as in woodland, or they may be aquatic,
may even survive in drier conditions, such as
on walls or roofs of houses

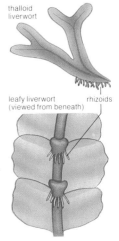

leafy liverwort   rhizoids
(viewed from beneath)

thallus (*n*) a general term for the plant body
which is not differentiated into root, stem or leaf
e.g. liverworts. **thalloid** (*adj*)

rhizoid (*n*) an elongate, single cell, such as in a
liverwort, or a multicellular (p. 9) thread, such
as in a moss, that anchors the gametophyte
(p. 177) to the substrate. It is not a true root.

protonema

cells

acrocarpous moss

pleurocarpous moss

foot[p](*n*) the lower part of the sporophyte (p. 171) generation of a bryophyte (↑) which remains embedded in the archegonium (p. 177).

capsule (*n*) (1) the end part of the sporophyte (p. 177) generation of liverworts or mosses which, at maturity, contains the spores (p. 178). (2) a dry fruit, such as that of the poppy, formed from two or more carpels (p. 179) which, during dehiscence (p. 185), open by a variety of slits or pores (p. 120) to release the seeds.

seta[p] (*n*) the stalk of the capsule (↑).

columella (*n*) the central, sterile tissue (p. 83) within the capsule (↑) of liverworts and mosses.

calyptra (*n*) a hood-like structure which covers the capsule (↑) of mosses until mature. It is the remains of the archegonium (p. 177).

operculum[p] (*n*) the lid of a moss capsule (↑) which is shed to reveal the peristome teeth (↓).

peristome teeth a ring of teeth at the tip of the capsule (↑) of mosses which open and close in response to varying levels of moisture – they open when dry and close when moist.

elater (*n*) a spindle-shaped body contained in the capsule (↑) of liverworts. Spiral thickenings change shape with varying moisture levels causing the elaters to flick spores (p. 178) from the capsule.

protonema (*n*) the branching filament (p. 181) that grows from the germinating spore (p. 178) of mosses. It develops buds which grow into the leafy gametophyte (p. 177) generation. Also known as **first thread protonemata** (*pl.*).

paraphyses

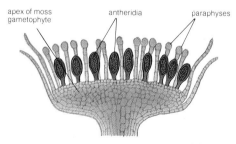

apex of moss gametophyte   antheridia   paraphyses

gemmae in thalloid liverworts: vegetative reproduction

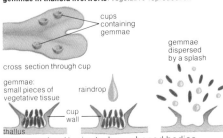

cups containing gemmae

cross section through cup

gemmae: small pieces of vegetative tissue

raindrop

cup wall

thallus

gemmae dispersed by a splash

liverwort sporophyte discharging spores

capsule walls

spores   elaters

elater with helical thickenings in cell wall

gemmae (*n.pl.*) minute, lens-shaped bodies produced by liverworts as a means of asexual reproduction (p. 173). **gemma** (*sing.*)

gemmae cup the receptacle or cup-shaped body on the upper surface of the gametophyte (p. 177) generation in liverworts which contains the gemmae (↑).

Pteridophyta (*n.pl.*) a division of the Plantae (p. 50) including the Lycopodiales (↓), Equisetales (↓) and the Filicales (↓) Pteridophytes have a well-developed vascular system (p. 127). They are widely distributed, especially in the tropics, and live mainly on land. There is alternation of generations (p.176) between gametophyte (p.177) and sporophyte (p. 177) phases in which the latter is the most prominent when the plant is differentiated into roots, stems, leaves and rhizomes (p. 174).

vascular cryptogams an alternative name for the Pteridophyta (↑), so called because there is a clear vascular system (p. 127) but no prominent organs of reproduction (p. 173), such as in the angiosperms (p. 57).

homosporous (*adj*) of plants with only a single kind of spore (p. 178) which gives rise to a hermaphrodite (p. 175) generation of gametophytes (p. 177). It occurs in some of the Pteridophyta (↑).

heterosporous (*adj*) of plants with two distinct kinds of spores (p. 178) which give rise to male and female gametophyte (p. 177) generations respectively It occurs in some Pteridophytes (↑) and is thought to represent an evolutionary (p. 208) step towards the production of seeds.

homospory and heterospory in vascular plants

homosporous
bryophytes
some pteridophytes (e.g. ferns)

heterosporous
some pteridophytes (e.g. club mosses)
gymnosperms
angiosperms

heterosporous plants the production of microspores

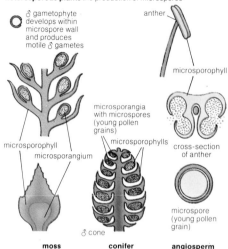

♂ gametophyte develops within microspore wall and produces motile ♂ gametes

anther

microsporophyll

microsporangia with microspores (young pollen grains)

microsporophyll

microsporangium

microsporophylls

cross-section of anther

microspore (young pollen grain)

♂ cone

moss　　**conifer**　　**angiosperm**

horsetail
Equisetales

strobilus

10 cm

1 cm

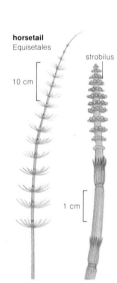

club moss Lycopodiales

spore-bearing shoots

strobilus (*n*) the reproductive (p. 173) structure of certain members of the Pteridophyta (↑). It consists of sporophylls (↓) on an axis. **strobili** (*pl.*).

coneᵖ = strobilus (↑).

Lycopodiales (*n.pl.*) club mosses. A division of the Pteridophyta (↑). They are an ancient group and even attained tree-like forms. They may be heterosporous (↑) or homosporous (↑) and bear densely packed small leaves on branched stems. They are evergreen.

sporophyll (*n*) a modified leaf that bears a sporangium (p. 178).

Equisetales (*n.pl.*) horsetails. A division of the Pteridophyta (↑). They are an ancient group and even attained tree-like forms. They are characterized by having whorls (p. 83) of small leaves on upright stems with strobili (↑) at the tips. They are homosporous (↑).

microphyll (*n*) a foliage leaf typical of the Lycopodiales (↑) and the Equisetales (↑) which may be very small and has a simple vascular system (p. 127) comprising a single vein running from the base to the apex.

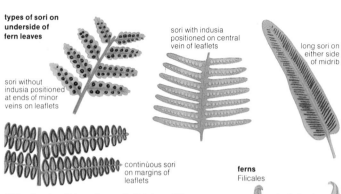

types of sori on underside of fern leaves

sori with indusia positioned on central vein of leaflets

long sori on either side of midrib

sori without indusia positioned at ends of minor veins on leaflets

continuous sori on margins of leaflets

ferns
Filicales

simple frond (megaphyll)

circinate vernation in young frond

compound frond (megaphyll)

Filicales (*n.pl.*) true ferns. A division of the Pteridophyta (p. 54). The plants are characterized by obvious frond-like leaves, often with sporangia (p. 178) on the undersides, and rhizomes (p. 174) underground. They are homosporous (p. 54).

megaphyll (*n*) a large frond-like foliage leaf with a branched system of veins. It is typical of the Filicales (↑).

frond (*n*) a large well-divided leaf typical of the Filicales (↑).

sorus (*n*) a reproductive (p. 173) organ made up of a group of sporangia (p. 178) which occur on the undersides of leaves in Filicales (↑).

indusium (*n*) a flap of tissue (p. 83) covering the sorus (↑). **indusia** (*pl.*).

annulus (*n*) an arc or ring of cells in the sporangia (p. 178) of Filicales (↑) which are involved in opening the sporangium on drying to release the spores (p. 178).

circinate vernation the way in which the young fronds (↑) of the Filicales (↑) occur rolled-up.

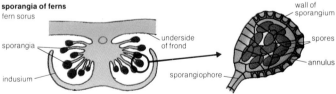

sporangia of ferns
fern sorus

sporangia

indusium

underside of frond

sporangiophore

wall of sporangium

spores

annulus

Spermatophyta (*n.pl.*) seed plants. A division of the Plantae (p. 54) including the Gymnospermae (↓) and the Angiospermae (↓). They are widely distributed and are the dominant plants on land today. The body is highly organized and differentiated into root, stem and leaf, and there is a well-developed vascular system (p. 127). They are heterosporous (p. 54) with a dominant sporophyte (p. 177) generation which is the plant itself. The male gametophyte (p. 177) is the pollen (p. 181) grain while the female gametophyte is the egg which becomes the seed after fertilization (p. 175).

conifer an example of a gymnosperm

Gymnospermae (*n.pl.*) a division of the Spermatophyta (↑) which includes trees and shrubs in which the seeds are naked and not enclosed in a fruit. Most have cones (p. 55).

Angiospermae (*n.pl.*) flowering plants. A division of the Spermatophyta (↑) which includes the dominant plants on land. They are highly differentiated and have microsporophylls (p. 178) and megasporophylls (p. 179) combined into true flowers as stamens (p. 181) and carpels (p. 179).

Dicotyledonae (*n*) dicotyledons. A class of the Angiospermae (↑) in which the seeds have two cotyledons (p. 168) or seed leaves, the leaves are net veined, the flower parts are usually in multiples of four or five, and the root system includes a tap root (p. 81) with lateral branches. An example of a dicotyledon is the buttercup.

dictotyledons some examples

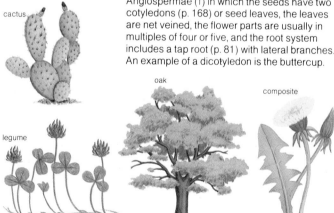

cactus

legume

oak

composite

Monocotyledonae (*n*) monocotyledons. A class of the Angiospermae (p. 57) in which the seeds have one cotyledon (p. 168) or seed leaf, the leaves are usually parallel veined, and the flower parts are usually in multiples of three. Typical monocotyledons are the grasses.

ephemeral (*adj*) of a plant in which the complete life cycle from germination (p. 168) to the production of seed and death is very short so that many generations of the plant may be completed within a single year.

annual (*adj*) of a plant that completes its whole life cycle from the germination (p. 168) of seeds to the production of the next crop of seeds, followed by the death of the plant, within one year.

biennial (*adj*) of a plant that completes its whole life cycle from the germination (p. 168) of seeds to the production of the next crop of seeds, followed by the death of the plant, within two years. During the first year, the plant produces foliage and photosynthesizes (p. 93) to provide an energy store for the reproductive (p. 173) activities of the second year.

perennial (*adj*) of a plant which survives for a number of years and may or may not reproduce (p. 173) within the first year.

monocotyledons
some examples

grass

palm

orchid

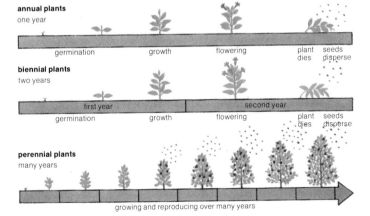

annual plants
one year

germination · growth · flowering · plant dies · seeds disperse

biennial plants
two years

first year · second year
germination · growth · flowering · plant dies · seeds disperse

perennial plants
many years

growing and reproducing over many years

two types of tree

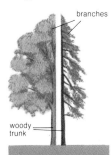

branches

woody trunk

deciduous with sympodial branching

evergreen with monopodial branching

deciduous (*adj*) (1) of a plant that sheds its leaves periodically, in accordance with the season, so that water loss by transpiration (p. 120) is reduced during periods of very dry or cold weather when water is in short supply. (2) in animals, of dentition (p. 104), for example, milk teeth (p. 104) which are shed and replaced by adult teeth.

evergreen (*adj*) of a plant that bears leaves throughout the year and which has adaptations, such as a leathery cuticle (p. 83) or needle-like leaves as in the gymnosperms (p. 57), conifers, to reduce water losses.

herbaceous (*adj*) of a perennial (↑) plant in which the foliage dies back each year while the plant survives as, for example, a bulb (p. 174), corm (p. 174), or tuber (p. 174). Herbaceous plants have no wood in their stems or roots.

tree (*n*) a woody, perennial (↑) plant which usually reaches a height of greater than 4 to 6 metres (13 to 20 feet) and which has a single stem from which branches grow at some distance from the ground level.

sapling (*n*) a young tree.

shrub (*n*) a woody, perennial (↑) plant which is smaller than a tree and from which branches grow quite close to ground level.

climber (*n*) a plant which though rooted in the ground uses other plants to support itself. Climbers use long coiled threadlike tendrils, suckers or adventitious roots (p. 81) to hold on to other plants and sometimes they twist around their stems.

foliage (*n*) all the leaves of a plant together. **foliar** (*adj*).

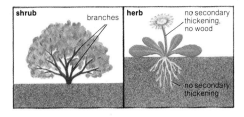

shrub branches

herb no secondary thickening, no wood

no secondary thickening

zoology (*n*) the study or science of animals or animal life.

Metazoa (*n*) a term used to describe all those truly multicellular (p. 9) animals as opposed to those animals which belong to the Protozoa (p. 44).

Coelenterata (*n*) a phylum of multicellular (p. 9) invertebrate (p. 75), aquatic and usually marine animals which includes the corals (↓) and jellyfishes. The body is radially symmetrical (↓) and consists of a simple body cavity which opens to the exterior by a mouth which is surrounded by a ring of tentacles (p. 71) that may have stinging cells or nematoblasts and are used for trapping prey and for defence. The body wall consists of an endoderm (p. 166) and an ectoderm (p. 166) separated by a jelly-like mesogloea. Reproduction (p. 173) takes place sexually, and asexually by budding (p. 173).

tissue grade the state of organization of animal cells into different types of tissue (p. 83) for different functions, such as muscular (p. 143) tissue and nervous tissue (p. 91) leading to greater co-ordination of activities such as response and locomotion (p. 143).

symmetrical (*adj*) of structures whose parts are arranged equally and regularly on either side of a line or plane (bilateral symmetry (p. 62)) or round a central point (radial symmetry (↓)).

asymmetrical (*adj*) not symmetrical (↑).

radial symmetry the condition in which the form of an organism is such that its structures radiate from a central point so that, if a cross section is made through any diameter, one half will be a mirror image of the other.

diploblastic (*adj*) of an animal whose body wall is composed of two layers, an endoderm (p. 166) and an ectoderm (p. 166), separated by a jelly-like mesogloea.

enteron (*n*) a sac-like body cavity which functions as a digestive (p. 98) tract or gut (p. 98).

planula larva the small, ciliated (p. 12) larva (p. 165) of a member of the Coelenterata (↑) which results from sexual reproduction (p. 173) and which swims to a suitable site before settling and growing into a polyp (↓).

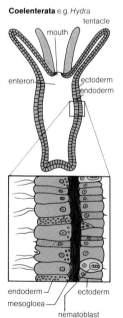

Coelenterata e.g. *Hydra*

tentacle

mouth

enteron

ectoderm

endoderm

endoderm

mesogloea

ectoderm

nematoblast

jellyfish
a scyphozoan

Physalia

pneumatophore — sea level
— gonozoid
— small dactylozoid
— gastrozoid
— fishing dactylozoid

Hydrozoa (*n*) a class of colonial and mainly marine Coelenterata (↑) in which alternation of generations (p. 176) is typical to give free-swimming medusae (↓) that reproduce (p. 173) sexually giving rise to sedentary polyps (↓) which reproduce asexually by budding (p. 173).

polyp (*n*) the sedentary stage in the life cycle of Coelenterata (↑) in which the body is tubular and surrounded by the tentacles (p. 71) at one end while attached to the substrate at the other. It reproduces (p. 173) asexually by budding (p. 173).

medusa (*n*) the free-swimming stage in the life cycle of Coelenterata (↑) in which the body is usually bell-shaped and surrounded by tentacles (p. 71) at one end. It reproduces (p. 173) sexually. **medusae** (*pl.*).

Scyphozoa (*n*) a class of the Coelenterata (↑) which comprises the jellyfishes and which may have no polyp (↑) form. The tentacles (p. 71) surround the mouth and bear stinging hairs.

Anthozoa (*n*) a class of marine Coelenterata (↑), including the sea anemones and corals (↓), in which the medusa (↑) stage is absent. The enteron (↑) is divided by vertical walls or septa and the animals may be colonial or solitary.

coral (*n*) any of the members of the Anthozoa (↑) which are today all colonial and in which the polyp (↑) is contained by a jelly-like, horny, or calcareous (containing $CaCO_3$) matrix (p. 88).

a single polyp
from the edge of a coral colony

peristome

mouth

tentacle

pharynx seen through body wall

column of body

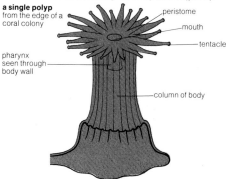

Platyhelminthes (*n*) a phylum of multicellular
(p. 9) invertebrate (p. 75) animals which
includes the flatworms (↓). The body is
bilaterally symmetrical (↓) and worm-like, and
consists of a single opening to the gut which is
often branched. There is no coelom (p. 167) or
vascular system (p. 127). The body wall
consists of an ectoderm (p. 166), mesoderm
(p. 167), and endoderm (p. 166). They are
usually hermaphrodite (p. 175).

flatworm (*n*) any of the members of the
Platyhelminthes (↑) which have a flattened
shape from above downwards that allows the
oxygen used in respiration (p. 112) to diffuse
into all parts of the body. There are three
groups which include the mainly marine
flatworms proper, the parasitic (p. 92)
tapeworms, and the parasitic flukes.

Platyhelminthes e.g. *Planaria*

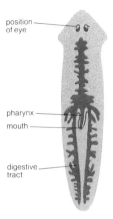

position
of eye

pharynx

mouth

digestive
tract

Planaria transverse section of body

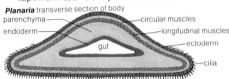

parenchyma

endoderm

gut

circular muscles

longitudinal muscles

ectoderm

cilia

triploblastic (*adj*) of those animals, such as the
Platyhelminthes (↑), in which the body wall
consists of three layers, the ectoderm (p. 166),
the mesoderm (p. 167), which is formed from
cells which have moved from the surface layer,
and the endoderm (p. 166).

bilateral symmetry the condition in which one
half of the organism, from a section taken down
its long axis, is a mirror image of the other half.

acoelomate (*adj*) of those animals without a
coelom (p. 167) e.g. Platyhelminthes (↑).

flame cell one of a number of cup-shaped cells
that occur in animals, such as the
Platyhelminthes (↑), which, by the beating of
their cilia (p. 12), draw fluid waste products into
their cavity, and then to the exterior.

sucker (*n*) an organ of attachment, for example,
in parasitic (p. 92) Platyhelminthes (↑) an
adaptation of the pharynx (p. 99) is used to
attach the organism to the host (p. 110).

Turbellaria (*n*) a class of the Platyhelminthes (↑) which includes free-living, mainly aquatic flatworms (↑) with a ciliated (p. 12) ectoderm (p. 166).

Planaria (*n*) a genus (p. 40) of the Turbellaria (↑) which includes freshwater forms that have numerous cilia (p. 12) on the underside that assist in locomotion (p. 143), respiration (p. 112) and direct food particles into the mouth.

Trematoda (*n*) a class of the Platyhelminthes (↑) which includes internal parasites (p. 110), such as the flukes, that have a complex life cycle including more than one host (p. 110), usually a vertebrate (p. 74) and an invertebrate (p. 75). They have suckers (↑), a branched gut (p. 98), and a thickened cuticle (p. 145) to resist digestion (p. 98) by the host.

bilharzia (*n*) a disease of humans living especially in Africa which is caused by a liver fluke that spends part of its life in freshwater snails, which are eaten by fish and then by humans. It enters the liver (p. 103) from the gut (p. 98) along the bile duct (p. 101).

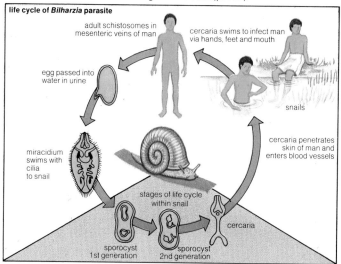

life cycle of *Bilharzia* parasite

adult schistosomes in mesenteric veins of man

cercaria swims to infect man via hands, feet and mouth

egg passed into water in urine

snails

miracidium swims with cilia to snail

cercaria penetrates skin of man and enters blood vessels

stages of life cycle within snail

cercaria

sporocyst 1st generation

sporocyst 2nd generation

Cestoda (n) a class of the Platyhelminthes (p. 62) which includes the internal parasites (p. 110), the tapeworm, that have a complex life cycle including more than one host (p. 110), both usually vertebrates (p. 74). They are armed with suckers as well as powerful grappling hooks on the head for attachment to the wall of the host's gut (p. 98). The body is divided into sections and has a tough cuticle (p. 145) to prevent digestion (p. 98) by the host.

Nematoda (n) a phylum of multicellular (p. 9) invertebrate (p. 75) animals which includes the roundworms (↓). The phylum includes terrestrial (p. 219), aquatic, and parasitic (p. 92) forms which have no cilia (p. 12), and an alimentary canal (p. 98) with a mouth and an anus (p. 103).

roundworm (n) any of the members of the Nematoda (↑) which have a characteristic rounded, unsegmented body that tapers at each end. They move by lashing the whole body into s shapes. The sexes are usually separate and the females lay large numbers of eggs. They are able to withstand adverse conditions by secreting (p. 106) a protective coat around the body.

threadworm (n) = roundworm (↑).

pseudocoel (n) a fluid-filled body cavity between the digestive (p. 98) tract and the other organs of roundworms (↑).

Annelida (n) a phylum of multicellular (p. 9) invertebrate (p. 75), mainly free-living, and typically marine aquatic animals which includes the 'true' segmented worms (↓). They have a central nervous system (p. 149), a thin cuticle (p. 145), and bristle-like chaetae (↓) on the body.

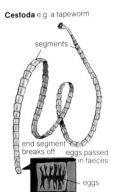

Cestoda e.g. a tapeworm

segments

end segment
breaks off   eggs passed
in faeces

eggs

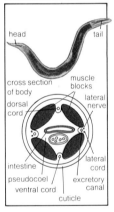

an adult roundworm

head   tail

cross section   muscle
of body   blocks

dorsal   lateral
cord   nerve

intestine   lateral
                        cord

pseudocoel   excretory
ventral cord canal
              cuticle

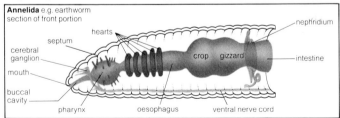

Annelida e.g. earthworm
section of front portion

nephridium

hearts

septum

cerebral
ganglion   crop   gizzard   intestine

mouth

buccal
cavity

pharynx   oesophagus   ventral nerve cord

segmented worm any of the members of the
Annelida (↑). They have a body which is divided
into obvious ring-like segments. Digestion
(p. 98) takes place in a simple, tube-like gut
(p. 98) which runs from the mouth at the front to
the anus (p. 103) at the rear. Between the body
wall and the gut is a fluid-filled coelom (p. 167).
They are hermaphrodite (p. 175).

nephridium (*n*) an organ which is used for excretion
(p. 134) in some invertebrates (p. 75), e.g. Annelida
(↑). It consists of a tube which opens to the exterior
at one end and, at the other, to flame cells (p. 62)
or to the coelom (p. 167). **nephridia** (*pl.*).

chaeta (*n*) one of a number of bristle-like
structures, composed of chitin (p. 49) which
are present, and arranged segmentally, along
the outside of the bodies of Annelida (↑). They
may assist in locomotion (p. 143) and, for
example, help earthworms (p. 66) to grip the
soil in which they live. **chaetae** (*pl.*).

chaeta transverse section
of earthworm body wall

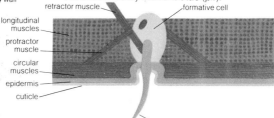

retractor muscle

formative cell

longitudinal
muscles

protractor
muscle

circular
muscles

epidermis

cuticle

chaeta

seta[a] (*n*) = chaeta (↑).

cerebral ganglion one of the pair of solid strands
of nervous tissue (p. 91) which runs ventrally
and forms part of the central nervous system
(p. 149) in Annelida (↑) and other invertebrates
(p. 75) and to which the ganglia (p. 155) are
connected segmentally.

nerve cord = cerebral ganglion (↑).

Polychaeta (*n*) a class of marine Annelida (↑)
which includes the bristleworms, ragworms,
lugworms etc, that have many chaetae (↑). The
sexes are usually separate.

parapodium (*n*) in members of the Polychaetae (↑),
one of many extensions of the body wall on which
the chaetae (↑) are found. **parapodia** (*pl.*).

trochosphere larva the larva (p. 165) of Annelida
(p. 64) and some other groups of invertebrates
(p. 75) which may be related through evolution
(p. 208). It is free-swimming, planktonic (p. 227),
and covered with cilia (p. 12), especially around
the mouth which leads to the digestive (p. 98)
tract and anus (p. 103).

Hirudinea (*n*) a class of ectoparasitic (p. 110),
freshwater Annelida (p. 64) which includes the
leeches (↓). They have no chaetae (p. 65) or
parapodia (p. 65) and are hermaphrodite (p. 125)

leech   sucker

leech (*n*) any of the Hirudinea (↑) which are
flattened and have a small sucker (p. 62) at
the front end and a larger, more obvious one at
the hind end. Some are carnivorous (p. 109)
but most are parasitic (p. 92), feeding on the
blood of their host (p. 110).

Oligochaeta (*n*) a class of mainly terrestrial
(p. 219) and freshwater Annelida (p. 64) which
includes the earthworms (↓). They have few
chaetae (p. 65), no parapodia (p. 65) and are
hermaphrodite (p. 175).

earthworm (*n*) any of the members of the
Oligochaeta (↑) which comprise the genus (p. 40)
Lumbricus. They live by burrowing in the soil
and digesting (p. 98) any organic matter in it
and are important in improving the structure of
the soil. They have a few chaetae (p. 65) and
secrete (p. 106) mucus (p. 99) from their skin.

clitellum (*n*) a saddle-like swelling of the
epidermis (p. 131) in earthworms (↑) which
binds the worms together during copulation
(p. 191) and then secretes (p. 106) the cocoon (↓).

cocoon (*n*) the protective covering , e.g. for the
eggs of an earthworm (↑) which is secreted
(p. 106) by the clitellum (↑).

clitellum
copulation between earthworms

clitellum

parts of an insect

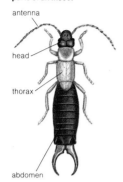

antenna

head

thorax

abdomen

Arthropoda (n) a phylum of multicellular (p. 9), invertebrate (p. 75) animals that occupy aerial, terrestrial (p. 219), freshwater and marine environments (p. 218) and make up some 80 per cent of known animal life. Their bodies are highly organized: a *head*, segments at the forward end, containing the organs for feeding and sensation as well as the brain (p. 155); a *thorax*, segments between the head and the abdomen which bear the jointed appendages (↓), and when present, the wings; and an *abdomen*, segments at the rear end. They are bilaterally symmetrical (p. 62), and protected by a tough exoskeleton (p. 145) which is segmented for mobility. They often have compound eyes. Growth takes place by ecdysis (p. 165). Each segment usually bears a pair of jointed appendages. The coelom (p. 167) is small and the main body cavity is a haemocoel (p. 68) containing a tube which is able to contract and function as a heart (p. 124). There is a nerve cord (p. 65) lying below the gut (p. 98) connected to paired ganglia (p. 155) for each segment. The best-known members of this phylum are the insects (p. 69) and spiders (p.70).

metameric segmentation the condition in which the body of an animal, especially certain invertebrates (p. 75) such as Annelida (p. 64) and Arthropoda (↑), is divided into a series of clearly definable units which are essentially similar to one another and repeat their patterns of blood vessels (p. 127), organs of excretion (p. 134) and respiration (p. 112), nerves (p. 149) etc. In the Arthropoda, the similarity between the units is reduced especially at the head end.

appendage (n) any relatively large protuberance or projection from the main body of an organism.

jointed appendage any one of the projections from the body of an arthropod (↑) which is divided into a number of segments, seven in insects (p. 69), and which are hinged between the segments to allow for articulation (bending) in different planes. The appendages are modified for different functions e.g. locomotion (p. 143), feeding, reproduction (p. 173), and respiration (p. 112).

antenna (*n*) one of the pair of highly mobile, thread-like, jointed appendages (p. 67) which occur on the head of an arthropod (p. 67) which are used mainly for touch and smell although, in some members of the group, they may assist in locomotion (p.143). **antennae** (*pl.*).

haemocoel (*n*) a blood-filled (p. 90) cavity which forms the main body cavity in arthropods (p. 67). The coelom (p. 167) is reduced to cavities surrounding the gonads (p. 187) etc while the haemocoel is essentially an expanded part of the blood system.

Crustacea   main groups

Crustacea (*n*) a class of the Arthropoda (p. 67), which includes the aquatic shrimps and crabs and the terrestrial (p. 219) woodlice. Typically, the body is divided into a head with two pairs of antennae (↑) and compound eyes, a thorax (p.115), and an abdomen (p. 116). The exoskeleton (p. 145) may be hardened by calcite ($CaCO_3$).

copepod (*n*) any of the group of small, aquatic crustaceans (↑) which form an important part of the marine plankton (p. 227). They have no carapace (↓) or compound eyes and the first pair of appendages (p. 67) on the head are modified for filter feeding (p. 108) while the six pairs on the thorax (p. 115) are used for swimming. The abdomen (p. 116) has no appendages.

copepod

isopod (*n*) any of the flattened, terrestrial (p. 219), freshwater, marine and often parasitic (p. 92) members of the Crustacea (↑) which have no carapace (↓) e.g. woodlice and shore slaters.

decapod (*n*) any of the terrestrial (p. 219), freshwater and mainly marine members of the Crustacea (↑) e.g. the highly specialized crabs, lobsters and prawns. They often have an elongated abdomen (p. 116) which ends in a tail that enables them to escape predation (p. 220) by swimming rapidly backwards. The head and thorax (p. 116) may be fused and protected by a carapace (↓). There are five pairs of jointed appendages (p. 67) on the thorax which are used in locomotion (p. 143) and three pairs used in feeding. One or two of the pairs of the legs may bear pincers which are used in courtship and for defence.

isopod

decapod

Myriapoda

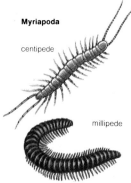

centipede

millipede

**internal structure of
an insect**

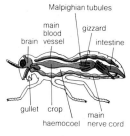

Malpighian tubules

main
blood
vessel

brain

gizzard

intestine

gullet   crop

haemocoel   main
nerve cord

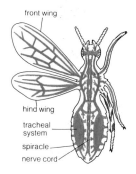

front wing

hind wing

tracheal
system

spiracle

nerve cord

carapace (n) a toughened, shield-like part of the exoskeleton (p. 145) which protects the back and sides of the head and thorax (p. 115) in some arthropods (p. 67), such as the crabs.

crayfish (n) any of the relatively small, freshwater decapods (↑), which resemble and are related to the marine lobsters. It has an elongated carapace (↑) and a flexible abdomen (p. 116). The first of the five pairs of jointed appendages (p. 67) on the thorax (p. 115) are modified to form large pincers that are used for feeding and defence. The remaining four pairs are used for locomotion (p. 143).

Myriapoda (n) a class of terrestrial (p. 219) arthropods (p. 67), including centipedes (↓) and millipedes (↓), which have elongated bodies with many segments, each bearing one or more pairs of jointed appendages (p. 67). They have a definite head bearing antennae (↑) and mouth parts.

chilopod (n) any of the flattened, carnivorous (p. 109) myriapods (↑), including the centipedes (↓), which have one pair of legs on each segment of which the first pair contains poison glands (p. 87).

centipede (n) any of the chilopods (↑), but especially members of the genus (p. 40) *Lithobius*, the members of which live beneath stones.

millipede (n) any of the rounded, herbivorous (p. 109) myriapods (↑) which have four single segments at the front end of the body and numerous double segments each with two pairs of legs.

Insecta (n) the largest, most diverse (p. 213) and important class of the Arthropoda (p. 67), among which the majority are terrestrial (p. 219) or aerial. Insects include the majority of all the known animals with more than a million species (p. 40) described and perhaps another thirty million awaiting identification. The body is characteristically divided into a head with a pair of antennae (↑), compound and simple eyes and mouth parts adapted for different feeding methods; a thorax (p. 115) with three pairs of legs and, usually, two pairs of wings for flight; and a limbless abdomen (p. 116). They have a waterproof exoskeleton (p. 145) and a very efficient means of respiration (p. 112) by means of tracheae (p. 115).

metamorphosis (*n*) the process by which, under the control of hormones (p. 130) some animals, e.g. insects (p. 69) change rapidly in form from the larva (p. 165) into the adult with considerable destruction of the larval tissue (p. 83).

proboscis (*n*) an extension of the mouth parts of an insect (p. 69) which is used for feeding.

commissure (*n*) a nerve cord (p. 65) which connects the segmental ganglia (p. 155) in the arthropods (p. 67).

Arachnida (*n*) the class of mainly terrestrial (p.219) arthropods (p. 67), which includes the spiders (↓), scorpions and pseudoscorpions. The body is divided into two main regions, the prosoma (↓) and opisthosoma (↓), and there are four pairs of walking legs, a pair of chelicerae and a pair of pedipalps (↓). The head has simple eyes and no antennae (p. 68). Respiration (p. 112) is achieved by book lungs (↓).

prosoma (*n*) the front region of the body of an arachnid (↑) which is made up of the head and thorax (p. 115) fused together.

opisthosoma (*n*) the hind region or abdomen (p. 116) of an arachnid (↑).

spider (*n*) an arachnid (↑) in which the prosoma (↑) and opisthosoma (↑) are separated by a narrow waist and which has spinnerets (↓).

pedipalps (*n.pl.*) the second pair of jointed appendages (p. 67) on the prosoma (↑) of arachnids (↑). They may be used for seizing prey, adapted as antennae (p. 68), or used for purposes of copulation (p. 191).

book lung one of the organs of respiration (p. 112) in arachnids (↑). They are composed of many fine layers of tissue (p. 83), resembling a book, through which blood (p. 90) flows and absorbs (p. 81) oxygen.

web (*n*) the thin, silken material which is spun by spiders (↑) in a variety of forms and used to capture prey such as flying insects, as in a net.

spinneret (*n*) one of the pair of appendages (p. 67) which is situated on the opisthosoma (↑) of spiders (↑) and secretes (p. 106) a liquid that hardens into silk for making webs (↑), wrapping the cocoon (p. 66), or binding the prey.

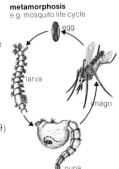

metamorphosis
e.g. mosquito life cycle

egg
larva
imago
pupa

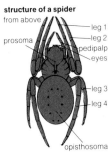

structure of a spider

from above

prosoma
leg 1
leg 2
pedipalp
eyes
leg 3
leg 4
opisthosoma

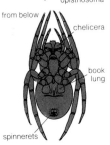

from below

chelicera
book lung
spinnerets

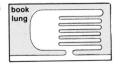

book lung

Mollusca

Gastropoda e.g. a snail

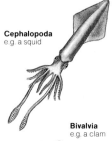

Cephalopoda
e.g. a squid

Bivalvia
e.g. a clam

adult snail and larva
(trochophore)

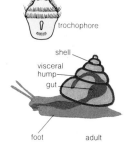

trochophore

shell
visceral hump
gut

foot   adult

Mollusca (*n*) a phylum of bilaterally symmetrical (p. 62), invertebrate (p. 75), multicellular (p. 9) animals that occupy terrestrial (p. 219), freshwater, and marine environments (p. 218) and include cockles, slugs, snails etc. They have soft, unsegmented bodies which are divided into a head region, a visceral hump (↓), and a foot (↓). In some groups of molluscs, the mantle (↓) secretes (p. 106) a hard shell. The coelom (p. 167) is reduced and there is a haemocoel (p. 68).

visceral hump the soft mass of tissue (p. 83) which makes up the bulk of a mollusc (↑) and which contains the main digestive (p. 98) system.

foot[a] (*n*) a soft, muscular (p. 143) development of the underside of the body of a mollusc (↑) which is used for locomotion (p. 143).

mantle (*n*) a fold of the body wall which covers the visceral hump (↑). In some molluscs (↑) it secretes (p. 106) a shell composed of calcium carbonate while, in others, it is folded to form a cavity which encloses the organs of respiration (p. 112).

radula (*n*) a tongue-like strip in most molluscs (↑) which is covered with horny teeth and is used to grind away food particles. As it is worn away, it is continuously replaced.

tentacle (*n*) a flexible appendage (p. 67). In cephalopods (p. 72), there are normally eight or ten extending from the foot (↑) which is incorporated into the head. Each tentacle bears many suckers and they are used for sense organs, for defence, and for grasping prey.

trochophore larva the free-swimming, ciliated (p. 12) larva (p. 165) of aquatic molluscs (↑).

Gastropoda (*n*) a class of terrestrial (p. 219), freshwater and marine molluscs (↑), including winkles, slugs, and snails, in which the visceral hump (↑) is coiled. This torsion (↓) of the visceral hump is reflected in the coiling of the shell. There is a muscular foot (↑) which is used in locomotion (p. 143), the eyes are on tentacles (↑), and gastropods feed using a radula (↑).

torsion (*n*) the act or condition of being twisted.

snail (*n*) any of the terrestrial (p. 219) members of the Gastropoda (p. 71), which have no gills (p. 113) but the mantle (p. 71) cavity functions as a lung (p. 115). This includes the slugs which lack the shell found in true snails.

bivalve e.g. razor clam

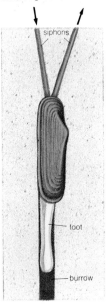

Bivalvia (*n*) a class of flattened freshwater and marine molluscs (p. 71) in which the mantle (p. 71) occurs in two parts and secretes (p. 106) a shell consisting of two hinged valves which may be pulled together by powerful muscles (p. 143). They have a poorly developed head and are filter feeders (p. 108). Some bivalves burrow into sand, mud, rock, or wood, some are attached to the substrate by strong threads, and others may be free swimming, propelling themselves backwards by forcibly opening and closing the valves.

mussel (*n*) any of a group of typical members of the Bivalvia (↑) which include freshwater and marine forms that have powerful muscles (p. 143) to clamp their valves tightly closed for protection. They are attached firmly to the substrate, such as rocks, by strong threads.

siphon (*n*) a tube, e.g. one of two tubes which protrude at the posterior end between the open valves of a bivalve (↑) mollusc (p. 71) and which form part of the system that circulates water through the mantle (p. 71) cavity for feeding and respiration (p. 112).

Cephalopoda (*n*) a class of marine molluscs (p. 71) with a well-developed head containing a complex brain (p. 155) and eyes. The head is surrounded by a ring of sucker-covered tentacles (p. 71) which is a modification of the foot (p. 71). There is a muscular siphon (↑) for respiration (p. 112) and the shell is much reduced and usually internal.

octopus (*n*) any of the members of the Cephalopoda (↑) with eight arm-like tentacles (p. 71) and a soft, ovally-shaped body.

octopus

Echinodermata

a starfish

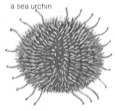

a sea urchin

a sea lily

Echinodermata (*n*) a phylum of radially symmetrical (p. 60) and usually five-rayed (↓), multicellular (p. 9), invertebrate (p. 75) animals that occupy marine environments (p. 218) and include the starfishes (↓) and sea urchins (p. 74). They have no head and a simple nervous system (p. 149). Part of the coelom (p. 167) is adapted to become a water vascular system (↓) which is unique to the group and connects with the tube feet (↓) which are used in locomotion (p. 143) and feeding. They have an internal skeleton (p. 143) of plates composed of calcite ($CaCO_3$) and most of them have spines.

spiny-skinned animal any of the members of the Echinodermata (↑) in which the ectoderm (p. 166) is covered with sharp, moveable, calcareous ($CaCO_3$) spines which connect with the calcareous ossicles (↓).

five-rayed radial symmetry radial symmetry (p. 60) which is typical of the Echinodermata (↑) in which there are five axes of symmetry.

tube foot any of the mobile, hollow, tube-like appendages (p. 67) which connect with the water vascular system (↓) and may end in suckers. They are used for locomotion (p. 143), feeding, and, in the sedentary sea-lilies, have cilia (p. 12) and are used for collecting food particles.

water vascular system a vascular system (p. 127) which is unique to the Echinodermata (↑) and consists of a series of canals containing sea water which, under pressure, operate the tube feet (↑).

madreporite (*n*) a sieve plate on the upper surface of echinoderms (↑) which is the opening of the water vascular system (↑) to the exterior.

calcareous ossicle any of the bone-like plates, made of calcium carbonate which make up the internal skeleton (p. 143) of the Echinodermata (↑).

starfish (*n*) any of the group of flattened, star-shaped Echinodermata (↑) which, typically, have five flexible arms radiating from the central disc which contains the main organs and the mouth on the underside. The arms have tube feet (↑) on the underside which are used for locomotion (p. 143) and for gripping prey. They usually live in the littoral (p. 219) zone.

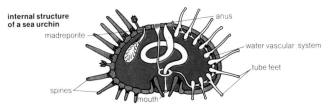

internal structure of a sea urchin — madreporite, anus, water vascular system, tube feet, spines, mouth

sea urchin any of the group of usually, globular,
heart-shaped, or disc-shaped Echinodermata (p. 73)
which have no arms and in which the calcareous
ossicles (p. 73) are fused together to form a rigid,
shell-like skeleton (p. 143) to which are
attached spines that can be moved by the water
vascular system (p. 73). They usually live on
or buried in the sea bed feeding on plants and
other debris through the mouth on the underside.

Chordata (*n*) a phylum of bilaterally symmetrical
(p. 60), invertebrate (↓) and vertebrate (↓)
multicellular (p. 9) animals, that includes humans
and other mammals (p. 80) and is characterized
by possessing a stiff, rod-like notochord (p. 167)
during some stage of their life cycle.

cranium (*n*) the part of the skeleton (p. 143),
composed of bone, of a vertebrate (↓) member
of the Chordata (↑) which is also referred to as
the skull and which contains the brain (p. 155).

vertebral column the part of the skeleton (p. 143)
of a vertebrate (↓) member of the Chordata (↑)
which is situated along the dorsal length of the
body, from the cranium (↑) to the tail (↓), and is
made of a linked chain of small bones or
cartilages (p. 90), the vertebrae. It is flexible
and allows for movement and locomotion
(p. 143). It replaces the notochord (p. 167) and
is a hollow column containing the spinal cord
(p. 154). Also known as **spine** or **backbone**.

visceral cleft one of the paired openings in the
pharynx (p. 99) which occur at some stage in the
life cycle of members of the Chordata (↑) and
persist in the aquatic species (p. 40). They lead from
the exterior to the gills (p. 113) and are involved
with filter feeding (p. 108) and gas exchange
(p. 112) as water is pumped through them.

vertebrate (*n*) an animal with a vertebral column (↑).

the three main groups of fish

Agnatha
e.g. lamprey

cartilaginous fish
e.g. shark

teleost fish
e.g. perch

invertebrate (*n*) an animal without a vertebral column (↑).

tail (*n*) an extension of the vertebral column (↑) which continues beyond the anus (p. 103) in most vertebrate (↑) members of the Chordata (↑.) It may be used for locomotion (p. 143), for balance and manoeuvrability (↓), or as a fifth limb.

manoeuvrability (*n*) the ability to make controlled changes of movement and direction.

Gnathostomata (*n*) a subphylum or superclass of the vertebrate (↑) Chordata (↑) which are characterized by the possession of a jaw (p. 105). The notochord (p. 167) is not retained throughout the life history.

Agnatha (*n*) a subphylum or superclass of the vertebrate (↑) Chordata (↑) which are characterized by having no jaw (p. 105). They are aquatic and primitive (p. 212).

Pisces the class of the Chordata (↑) which contains the fish. Fish are freshwater and marine animals with streamlined bodies that are usually covered with scales (p. 76). They have a powerful, finned (↓) tail (↑) which is used to propel them through the water, while their pairs of pelvic (↓) and pectoral (↓) fins are used for stability and manoeuvrability (↑). Gas exchange (p. 112) takes place in the gills (p. 113) and fish are exothermic (p. 130).

fin (*n*) a flattened, membraneous external organ on the body of a fish which usually occurs in pairs. It is used for steering, stability, and propulsion.

pectoral (*adj*) of the chest, e.g. the pectoral fins (↑) of a fish are attached to the shoulder and are used for steering up or down in the water and for counteracting pitching and rolling.

pelvic (*adj*) of the pelvic girdle (p. 147), e.g. the pelvic fins (↑) of a fish are attached to the pelvic girdle and are used for steering up or down and for counteracting pitching and rolling.

dorsal (*adj*) at, near or towards the back of an animal i.e. that part which is normally directed upwards (or backwards in humans).

ventral (*adj*) at, near or towards the part of an animal that is normally directed downwards (or forwards in humans).

scale (*n*) one of the many bony or horny plates which are made in the skin of fish and which may be above or beneath the skin. They overlap to form a protective and streamlined covering for the fish. Under the microscope (p. 9), it can be seen that they have a ring-like structure which represents the growth rate of the fish and can be used for aging purposes.

Chondrichthyes (*n*) a subclass of the Pisces (p. 75) which are entirely marine and include the sharks and rays. They are characterized by having an internal skeleton (p. 143) made of cartilage (p. 90) and are, therefore, also referred to as the cartilaginous fish. They have no swim bladder (↓) so that they sink if they cease moving.

cartilaginous fish = Chondrichthyes (↑).

Osteichthyes (*n*) a subclass of the Pisces (p. 75) which includes both freshwater and marine forms. They are characterized by having an internal skeleton (p. 143) and scales (↑) made from bone and are, therefore, referred to as bony fish. They possess a swim bladder (↓).

bony fish = Osteichthyes (↑).

teleost fish any of the main group of Osteichthyes (↑) in which the body tends to be laterally flattened and which have a swim bladder (↓) to adjust their buoyancy (↓). Their fins (p. 75) are composed of a thin, membraneous skin supported on bony rays. Their jaws (p. 105) are shortened so that the mouth can open widely and the visceral clefts (p. 74) are protected by a covering operculum (p. 113). The scales (↑) are thin, bony, and rounded. There is a wide variety of types of teleost fish and they occupy most aquatic environments (p. 218).

external features of a teleost fish

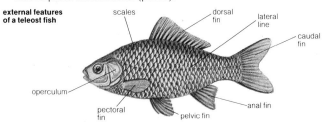

scales
dorsal fin
lateral line
caudal fin
operculum
pectoral fin
pelvic fin
anal fin

mermaid's purse

swim bladder a sac situated within the abdominal (p. 116) cavity of bony fish (↑). It contains a mixture of oxygen and nitrogen and oxygen can be pumped into it from the blood (p. 90) to increase the fish's buoyancy (↓) or vice versa so that the fish's depth can be controlled. It also functions as a sound detector and producer and, in lung fish, enables respiration (p. 112) out of water.

buoyancy (*n*) the ability to float in a liquid.

mermaid's purse the protective egg case which encloses the small number of eggs produced by cartilaginous fish (↑).

homocercal (*adj*) of a tail (p. 75), such as that of the teleost fish (↑), which is symmetrical (p. 60) in shape.

heterocercal (*adj*) of a tail (p. 75), such as that of the cartilaginous fish (↑), which is asymmetrically (p. 60) shaped such that the lower fin (p. 75) is larger than the upper fin to give the fish additional lift thereby compensating for the lack of a swim bladder (↑).

homocercal tail fin

tetrapod (*n*) any of the vertebrate (p. 74) members of the Chordata (p. 74), such as a mammal (p. 80), which have two pairs of limbs for support, locomotion (p. 143), etc.

pentadactyl (*adj*) of the limb of a tetrapod (↑) which terminates in five digits, although the digits may be reduced or fused together as adaptations to various modes of life.

heterocercal tail fin

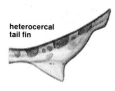

Amphibia (*n*) a class of primitive (p. 212) tetrapod (↑) chordates (p. 75), such as the frogs and toads, among which fertilization (p. 175) is external so that they must return to water to breed. Their larval (p. 165) forms are all aquatic and have gills (p. 113) but the majority of the adults are able to survive in damp conditions on land because they have a lung (p. 115) and are able to breathe air, respiring (p. 112) mainly through the thin, porous skin. Because of the thin skin, body fluids are easily lost. Like fish, they are exothermic (p. 130).

Amphibia e.g. salamander

salamanders (*n.pl.*) members of the order Urodela of the Amphibia (↑) which includes tailed amphibia. The order Urodela also includes newts.

**life cycle of
a frog**

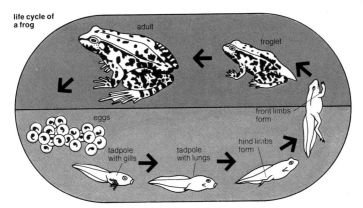

adult

froglet

eggs

front limbs
form

tadpole
with gills

tadpole
with lungs

hind limbs
form

Anura (*n*) the order of the Amphibia (↑) which includes the aquatic, tree-dwelling, or damp-loving frogs and the warty skinned toads which can survive in drier conditions. Their hind limbs are elongated and powerful for jumping and have webbed feet for swimming.

Reptilia (*n*) a class of the Chordata (p. 74) which includes the most primitive (p. 212) tetrapods (p. 77), such as the snakes and lizards, that are wholly adapted to terrestrial (p. 219) environments (p. 218) although some, such as the turtles, have returned to an aquatic existence. They are air breathing and possess a true lung (p. 115). Their skin is scaly so that it is resistant to loss of body fluids. Since fertilization (p. 175) is internal, there is no need to return to the water to breed, and they lay amniote (p. 191) eggs with a leathery skin from which the young develop without passing through a larval (p. 165) stage. Like the fish and amphibians (p. 77), they are exothermic (p. 130).

cleidoic (*adj*) of an egg, such as that of a reptile (↑), which has a waterproof covering or shell that is permeable to air.

Chelonia (*n*) an order of the Reptilia (↑), which contains the turtles and tortoises, that are characterized by the plates of bone, overlaid with further horny plates, that enclose the body.

**the four main groups of
Reptilia**

Chelonia
e.g. turtles

Ophidia
e.g. snakes

Lacertilia
e.g. lizards

Crocodylia
e.g. crocodiles

Squamata (*n*) an order of the Reptilia (↑), which contains the scaly skinned lizards and snakes.

Lacertilia (*n*) a suborder of the Squamata (↑) which contains the lizards. Most are truly tetrapod (p. 77) with a long tail (p. 75), have opening and closing eyelids, an eardrum (p. 158), and normal articulation of the jaws (p. 105).

Ophidia (*n*) a suborder of the Squamata (↑) which contains the snakes. They have elongated bodies with no limbs, no eardrum (p. 158) and no moveable eyelid. The jaws (p. 105) can be dislocated to allow a very wide gape so that large prey can be swallowed whole.

cloaca (*n*) the chamber which terminates the gut (p. 98) in all vertebrates (p. 74), other than the placental (p. 192) mammals (p. 80), and into which the contents of the alimentary canal (p. 98), the kidneys (p. 136), and the reproductive (p. 173) organs are discharged. There is a single opening leading to the exterior.

poison gland one of the modified salivary glands (p. 87) which may be present in some species (p. 40) of, for example, the Ophidia (↑), and which secrete (p. 106) toxic substances that may be, for example, injected into prey through the fangs.

Aves (*n*) a class of the Chordata (p. 74), which contains the birds. They are characterized by the possession of feathers (p. 147) for insulation and flight (p. 147) and other adaptations for flying. Although they are similar in many ways to the Reptilia (↑) from which they evolved (p. 208), they are endothermic (p. 130). There are some flightless species (p. 40). They lay amniote (p. 191) eggs with a calcareous ($CaCO_3$) shell.

internal organs of a bird

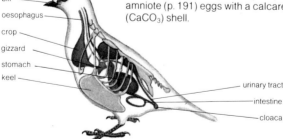

bill

oesophagus

crop

gizzard

stomach

keel

urinary tract

intestine

cloaca

bill (*n*) the horny structure which encloses the jaws (p. 105) of birds. It lacks teeth and may take a variety of forms adapted to different methods of feeding. Also known as **beak**.

keel (*n*) a bony projection of the sternum (p. 149) of birds to which the powerful pectoral (p. 75) muscles (p. 148) are attached for flight (p. 147).

air sac one of a number of thin-walled, bladder-like sacs in birds, which are connected to the lungs and which are present in the abdominal (p. 116) and thoracic (p. 115) cavities. They even penetrate into some of the bones of the skeleton (p. 143) to lighten the body of the bird without reducing its strength. The tracheae (p. 115) of some insects (p. 69) contain air sacs.

Mammalia (*n*) a class of the Chordata (p. 74) which contains all the mammals, e.g. dogs, cats and apes. They are endothermic (p. 130), have a glandular (p. 87) skin, and are covered with hair for insulation. They are characterized by possessing mammary glands which secrete (p.106) milk to feed the young. They possess heterodont dentition (p. 104), a secondary palate which enables them to eat and breathe at the same time and relatively large brains (p. 155).

Monotremata (*n*) a subclass of the Mammalia (↑) which includes the primitive (p. 212) spiny anteater and duck-billed platypus. They possess a cloaca (p. 79) and lay eggs. The young are transferred to a pouch and fed from milk which is secreted (p. 106) on to a groove in the abdomen (p. 116). They are covered with hair but have a relatively low body temperature. They have a poorly developed brain (p. 155).

Metatheria (*n*) a subclass of the Mammalia (↑) which includes the marsupial or pouched forms, such as the kangaroo. They are viviparous (p. 192) but the poorly developed live young are born after only a brief period of gestation (p. 192) and then transferred to a pouch where they are suckled and complete their growth.

Eutheria (*n*) a subclass of the Mammalia (↑) which contains the 'true' viviparous (p. 192), placental (p. 192) mammals.

the three main subclasses of mammals

Monotremata (monotremes)
e.g. duck-billed platypus

Metatheria (marsupials)
e.g. kangaroo

Eutheria (placental mammals)
e.g. elephant

tap root

fibrous roots

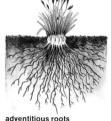

adventitious roots
on corm

contractile   adventitous
root   root

root cap L.S.

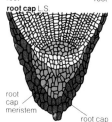

root
cap
meristem

root cap

anatomy (n) the study or science of the internal structure of animals and plants.

histology (n) the study or science of tissues (p. 83).

morphology (n) the study or science of the external structure and form of animals and plants without particular regard to their function and internal structure or anatomy (↑).

physiology (n) the study or science of the processes which take place in animals and plants.

root (n) the structure of a plant which anchors it firmly to the soil and which is responsible for the uptake of water containing mineral salts from the soil and passing them into the stem. A root may also function as a food store. Unlike underground stems, a root does not contain chlorophyll (p. 12) and cannot bear leaves or buds.

tap root a main, usually central root which may be clearly distinguished from the other roots in a root system.

adventitious root one of a number of roots which grow directly from the stem of the plant as in bulbs (p. 174), corms (p. 174), and rhizomes (p. 174) and which do not grow from a main root.

fibrous root one of a number of roots which grow at the same time as the germination (p. 168) of a plant, such as a grass, and from which other lateral roots grow.

root cap a layer of cells at the tip of a root which protects the growing point from abrasion and wear by soil particles etc.

root hair a fine, thin-walled, tube-shaped structure which grows out from the epidermis (p. 131) just behind the root tip and which is in intimate contact with the soil surrounding a root. It greatly increases the root's surface area for the uptake of water. Water is drawn into the root hair by osmosis (p. 118) because the root hairs and the piliferous layer (p. 82) contain fluid with a lower osmotic potential (p. 118) than the water in the soil.

absorb (v) to take in liquid through the surface. **absorption** (n).

piliferous layer a single layer of cells which surrounds the root tip and part of the root of a plant and from which the root hairs (p. 81) grow. The cells contain fluid with a lower osmotic potential (p. 118) than that of soil water so that it is the main region of absorption (p. 81) of the root.

leaf

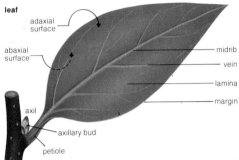

adaxial surface

abaxial surface

midrib

vein

lamina

margin

axil

axillary bud

petiole

leaf (*n*) a usually flattened, green structure which may or may not be joined to the stem of a plant by a stalk or petiole (↓). Its function is to make food for the plant in the form of carbohydrates (p. 17) by photosynthesis (p. 93).

vein[p] (*n*) one of a network of structures found in a leaf which provide support for the leaf and also transport the water, which is used during photosynthesis (p. 93), and organic (p. 15) solutes, into and out of the leaf tissue (↓).

lamina (*n*) the flat, thin, blade-shaped structure which comprises the major part of the foliage leaf.

petiole (*n*) the stalk or stem which may join the lamina (↑) to the stem of a plant.

midrib (*n*) the central or middle rib of a leaf which is an extension of the petiole (↑) into the leaf blade.

stem (*n*) that part of a plant which is usually erect and above ground and which bears the leaves, buds and flowers of the plant. Its function is to transport water and food throughout the plant, space out the leaves, and hold any flowers in a suitable position for pollination (p. 183).

a generalized flowering plant

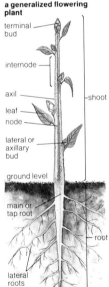

terminal bud

internode

axil

leaf

node

lateral or axillary bud

ground level

main or tap root

shoot

root

lateral roots

whorl

parenchyma cells
in cross section

cytoplasm

cell wall

plastid

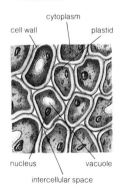

nucleus

vacuole

intercellular space

whorl (*n*) a group of three or more of the same organs arranged in a circle at the same level on a stem.

node (*n*) that part of the stem from which the leaves grow.

internode (*n*) the region of the stem between the nodes (↑).

bud (*n*) an undeveloped shoot which may develop into a flower or a new shoot. It consists of a short stem around which the immature leaves are folded and overlap. Buds may be at the tip of a shoot, when they are called terminal, or in the axils (↓) when they are called axillary.

shoot (*n*) the whole part of the plant which occurs above ground and which usually consists of a stem, leaves, buds and flowers.

axil (*n*) the angle between the upper side of a leaf and the stem on which the leaf is growing.

lenticel (*n*) a small raised gap or pore (p. 128) in the bark (p. 172) of a woody stem through which oxygen and carbon dioxide may pass.

tissue (*n*) a group of cells that perform a particular function in an organism.

vascular tissue a tissue (↑) which is specialized mainly for the transport of food and water throughout a plant, and is composed mainly of xylem (p. 84) and phloem (p. 84) together with sclerenchyma (p. 84) and parenchyma (↓).

ground tissue a tissue (↑), such as pith (p. 86) and cortex (p. 86), usually composed of parenchyma (↓), and which occupies all parts of the plant which do not contain the specialized tissue.

packing tissue = ground tissue (↑).

epidermal tissue a dermal (p. 131) tissue (↑) which forms a continuous outer skin over the surface of a plant. There are no spaces between the cells but it is penetrated by stomata (p. 120).

cuticle (*n*) the waterproof, waxy, or resinous outer surface of the epidermal tissue (↑) which occurs on the aerial (p. 219) parts of the plant.

parenchyma (*n*) a tissue (↑) which consists of rounded cells enclosed in a cellulose (p. 19) cell wall (p. 8) and containing air-filled intercellular (p. 110) spaces. Parenchyma supports the non-woody parts of a plant and also functions as storage tissue in the roots, stem and leaves.

collenchyma (*n*) a tissue (p. 83) composed of elongated cells in which the primary cell wall (p. 14) is unevenly thickened with cellulose (p. 19). Collenchyma tissue is specialized to provide support to actively growing parts of the plant which may also need to be flexible.

sclerenchyma (*n*) a tissue (p. 83) which has a secondary cell wall (p. 14) of lignin (p. 19) and which is composed of sclereids (↓) and fibres (↓). Its function is to provide support.

sclereid (*n*) one of the two types of cells which comprise the sclerenchyma (↑). It is not always easy to differentiate between a sclereid and a fibre (↓) although they are generally very little longer than they are broad and are the stone cells of the shells of nuts and the stones of fruits.

fibre[p] (*n*) one of the two types of cells which comprise the sclerenchyma (↑). They are elongated lignified (p. 19) cells with no living contents and provide great support.

xylem (*n*) a vascular tissue (p. 83) consisting of hollow cells with no living contents and additional supporting tissue (p. 83) including fibres (↑), sclereids (↑) and some parenchyma (p. 83). The cell walls (p. 8) are lignified (p. 19), the thickness varying in shape and extent. The two main cell types found in xylem are vessels (↓) and tracheids (↓).

tracheid (*n*) one of the two types of cells found in xylem (↑). A tracheid is elongated and has tapering ends and cross walls. Tracheids run parallel to the length of the organ which contains them. Each tracheid is connected to its neighbour by pairs of pits (p. 14) through which water can easily pass.

phloem (*n*) a vascular tissue (p. 83) which transports food throughout the plant (translocation (p. 122)). It contains sieve tubes (↓) and companion cells (↓), and, in some plants, may also contain other cells, such as parenchyma (p. 83) and fibres (↑).

sieve tube a column of thin-walled, elongated cells which are specialized to transport food materials through the plant.

collenchyma cells
in cross section

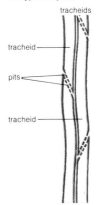

cellulose thickening

cell types in xylem

tracheids

tracheid

pits

tracheid

vessels

no end walls

spiral
thickening
of cell wall

phloem
(L.S.)

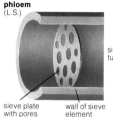

sieve plate with pores

wall of sieve element

sieve tube

position of xylem and phloem in young and old roots

pericycle and endodermis

cortex

phloem

xylem

young

wood

xylem

vascular cambium

phloem

cork

cork cambium

cork

old

position of xylem and phloem in young and old stems

epidermis

cortex

xylem

phloem

pith

young

phloem

xylem

vascular cambium

cork

cork cambium

cork

old

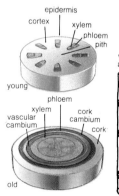

companion cells with nuclei

sieve elements

sieve plates

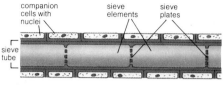

sieve plate the perforated end wall of a sieve tube (↑) through which strands of cytoplasm (p. 10) pass to connect the neighbouring cells.

companion cell a small, thin-walled cell containing dense cytoplasm (p. 10) and a well-defined nucleus (p. 13) situated alongside the sieve tube (↑) and which may aid the metabolism (p. 26) of the sieve tube.

vessel[P] (n) one of the two types of cells found in xylem (↑). Each vessel consists of a series of cells arranged into a tube-like form with no cross walls. It runs parallel to the length of the organ containing it and is found mainly in the angiosperms (p. 57). When mature, the vessel has no living contents, and has thick lignified (p. 19), walls for strength. There are several types of thickening: *annular* which has rings of lignin along the length of the cell; *spiral* which has a spiral or coil of lignin round the inner surface of the cell wall (p. 8); *scalariform* which has a ladder-like series of bars of lignin on the inner surface of the cell wall; *reticulate* which has a network of lignin over the inner surface of the cell wall; and *pitted* which has lignin over the whole inner surface of the cell wall except for many small pits (p. 14) or pores (p. 120).

vessels types of thickening

annular   spiral   scalariform   reticulate   pitted

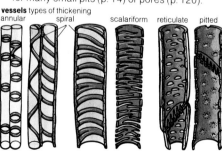

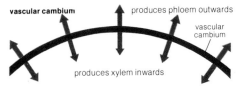

cambium (*n*) the layer of narrow, thin-walled cells which are situated between the xylem (p. 84) and phloem (p. 84) and give rise by division to secondary xylem (p. 172) and secondary phloem (p. 172). The cambium does not lose its ability to make new cells and is responsible for lateral growth in plants.

secondary tissue the additional tissue (p. 83) formed by the cambium (↑) leading to an increase in the lateral dimensions of the stem or root of a plant.

stele (*n*) the core or bundle of vascular tissue (p. 83) in the centre of the roots and stems of plants.

exodermis (*n*) the outer layer of thickened cells which may replace the epidermis (p. 131) in the older parts of roots.

endodermis (*n*) the layer of cells surrounding the stele (↑) on the innermost part of the cortex (↓) of a root.

cortex (*n*) the tissue (p. 83) usually of parenchyma (p. 83), which occurs in the stems and roots of plants between the stele (↑) and the epidermis (p. 131). It tends to make the stem more rigid.

pith (*n*) the central core of the stem composed of parenchyma (p. 83) tissue (p. 83) and found within the stele (↑).

medullary ray one of a number of plates of parenchyma (p. 83) cells which are arranged radially and pass from the pith (↑) to the cortex (↑) or terminate in secondary xylem (p. 84) and phloem (p. 84).

pericycle (*n*) the outermost layer of the stele (↑) with the endodermis (↑) and composed of parenchyma tissue (p. 83).

mesophyll (*n*) the tissue (p. 83) which lies between the epidermal (p. 131) layers of a leaf lamina (p. 82) and is involved in photosynthesis (p. 93).

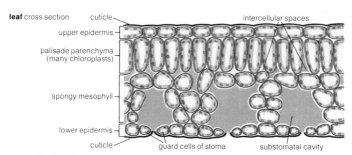

leaf cross section
cuticle
upper epidermis
palisade parenchyma
(many chloroplasts)
spongy mesophyll
lower epidermis
cuticle
intercellular spaces
guard cells of stoma
substomatal cavity

types of epithelium

pavement
(squamous)
nucleus
cytoplasm
basement membrane

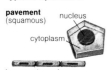

columnar

ciliated

glandular
goblet cell
secreting
mucus

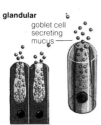

palisade mesophyll the mesophyll (↑) composed of cylindrical cells at right-angles to the leaf surface and situated just below the upper endodermis (↑). It contains numerous chloroplasts (p.12) and is concerned with photosynthesis (p.93).

spongy mesophyll the mesophyll (↑) composed of loosely and randomly arranged cells with few chloroplasts (p. 12) and large air spaces which are connected with the atmosphere through the stomata (p. 120).

epithelium (*n*) an animal tissue (p. 83) composed of a sheet of cells which are densely packed and which covers a surface or lines a cavity.

endothelium (*n*) the epithelium (↑) which lines the heart (p. 124) and blood vessels (p. 127).

basement membrane a membrane (p. 14) composed of a thin layer of cement to which one of the cells of the epithelium (↑) is fixed.

ciliated epithelium an epithelium (↑) bearing cilia (p.12) found in the trachea (p.115) and bronchi (p.116).

glandular epithelium an epithelium (↑) which is specialized to form secretory (p. 106) glands (↓).

goblet cell a wine-glass-shaped cell which secretes (p. 106) mucus (p. 99) on to the outside of columnar epithelium (↑) to protect it.

gland (*n*) an organ which secretes (p. 106) chemicals to the outside. **glandular** (*adj*).

compound epithelium an epithelium (↑) which is made up from more than one layer of cells with columnar cells attached to the basal membrane (p. 14) and squamous (flattened) cells furthest from it. It is found at areas of stress such as the epidermis (p. 131) of skin.

stratified epithelium = compound epithelium (p.87).

transitional epithelium a stratified epithelium (↑) that is also capable of stretching, and found in areas such as the bladder (p. 135).

connective tissue the tissue (p. 83) that functions for support or packing purposes in animals. It has a few quite small cells with greater amounts of intercellular (p. 110) or matrix (↓) material.

matrix (*n*) the intercellular (p. 110) ground substance in which cells are contained.

areolar tissue a connective tissue (↑) which surrounds and connects organs. It is composed of collagen (↓) and elastic fibres (↓) in an amorphous matrix (↑).

fibroblast (*n*) an irregularly shaped but often elongated and flattened cell which functions in the production of collagen (↓).

mast cell a cell present in the matrix (↑) of areolar tissue (↑) which produces anticoagulant (p. 128) substances and is also found in the endothelium (p. 87) of blood vessels (p. 127).

macrophage (*n*) a large cell found widely in animals but particularly in the connective tissue (↑). Macrophages wander freely through the tissue and in the lymph nodes (p. 128) by amoeboid movement (p. 44) and destroy harmful bacteria (p. 42) by engulfing them as well as helping to repair any damage to tissue (p. 83).

collagen fibre a non-elastic fibre with high tensile strength found in connective tissue (↑), particularly in tendons (p. 146), skin and skeletal (p. 145) material. Also known as a **white fibre**.

elastic fibre a highly elastic fibre (p. 143) found in connective tissue (↑), particularly in ligaments (p. 146) and organs, such as lungs (p. 115). Also known as **yellow fibre**.

adipose tissue a connective tissue (↑) similar to areolar tissue (↑) but containing closely packed fat cells and found under the skin and associated with certain organs to provide insulation, protection, and to store energy.

bone (*n*) a hard connective tissue (↑) composed of osteoblasts (p. 90) in a matrix (↑) made up of collagen fibres (p. 88) and calcium phosphate. It makes up the majority of the skeleton (p. 145).

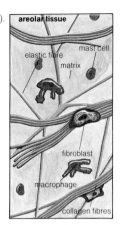

areolar tissue

mast cell
elastic fibre
matrix

fibroblast

macrophage

collagen fibres

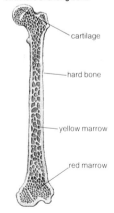

structure of a long bone

cartilage

hard bone

yellow marrow

red marrow

compact bone bone in which the Haversian canals (↓) are densely packed.

periosteum (*n*) connective tissue (↑) surrounding the bone and containing osteoblasts (p. 90) as well as many collagen fibres (↑) making it tough. The muscles (p. 143) and ligaments (p. 146) are attached to the periosteum.

Haversian canal a canal running along the length of bone and containing the nerves (p. 149) and blood vessels (p. 127) as well as the lymph vessels (p. 128) which secrete (p. 106) the osteocytes (p. 90).

Haversian system the system of Haversian canals (↑) surrounded by rings of bone and which connect with the surface of the bone and with its marrow (p. 90).

canaliculus (*n*) one of the fine canals linking the lacunae (↓) and containing the branches of the osteocytes (p. 90).

endosteum (*n*) a thin layer of connective tissue (↑) within a bone next to the cavity containing the marrow (p. 90).

lacuna (*n*) one of the spaces between the bone lamellae (↓) in which the osteoblasts (p. 90) are found. **lacunae** (*pl.*).

bone lamellae ring-like layers of calcified matrix (↑) in bone and surrounding the Haversian canals (↑).

compact bone
transverse section

periosteum

Haversian system

Haversian canal
bone lamellae
osteoblast { canaliculus
lacunae

endosteum

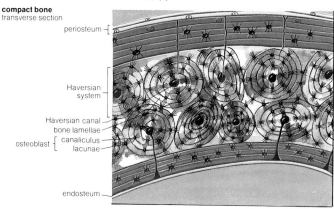

chondroblast (*n*) a cell which occurs in cartilage (p. 90) and secretes (p. 106) the matrix (p. 88) of clear chondrin (↓).

cartilage (*n*) a skeletal (p. 145) tissue (p. 83) composed of chondroblasts (↑) in a matrix (p. 88) of clear chondrin (↓). There are also many collagen fibres (p. 88) contained within it.

chondrin (*n*) a bluish-white clear gelatinous material which forms the ground substance of cartilage (↑). Chondrin is elastic.

hyaline cartilage cartilage (↑) which contains collagen fibres (p. 88) and which forms the embryonic (p. 166) skeleton (p. 145).

osteoblast (*n*) a cell present in the hyaline cartilage (↑) which is responsible for the laying down of the calcified matrix (p. 88) of bone.

osteocyte (*n*) an osteoblast (↑) which has become incorporated in the bone during its formation and has stopped dividing.

spongy bone bone which contains a network of bone lamellae (p. 89) surrounding irregularly placed lacunae (p. 89) containing red marrow (↓).

epiphysis (*n*) the end of the limb (p. 147) bone in mammals (p. 80) which enters and takes part in the joint (p. 146).

marrow (*n*) the soft, fatty tissue (p. 83) which is present in some bones and which produces the white blood cells (↓).

blood (*n*) the specialized fluid in animals which is found in vessels (p. 127) contained within endothelial (p. 87) walls and which may contain a pigment (p. 126) used in the transport of respiratory (p. 112) gases as well as transporting food and other materials throughout the body.

plasma (*n*) the clear, almost colourless fluid part of the blood (↑) which carries the white blood cells (↓), the red blood cells (↓) and the platelets (p. 128). It consists of 90 per cent water and 10 per cent other organic (p. 15) and inorganic (p. 15) compounds.

serum (*n*) the clear, pale-yellow fluid which remains after blood (↑) has clotted (p. 129) and consists essentially of plasma (↑) without the clotting agents.

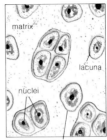

cartilage   chondroblast
(cartilage cell)

blood cells

**red blood cell
or erythrocyte**

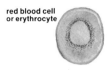

white blood cells

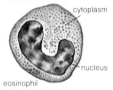
cytoplasm

nucleus

eosinophil

basophil

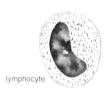

lymphocyte

nucleus   cytoplasm

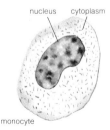

monocyte

red blood cell a blood cell which contains the respiratory (p. 112) pigment (p. 126), such as haemoglobin (p. 126).

erythrocyte (*n*) = red blood cell (↑).

white blood cell a blood cell which contains no respiratory (p. 112) pigment (p. 126). White blood cells are important in defending the body against disease because they are able to engulf bacteria (p. 42) as well as producing antibodies (p. 233).

leucocyte (*n*) = white blood cell (↑).

polymorphonuclear leucocyte a white blood cell (↑) with a dark staining, lobed nucleus (p. 13) and granular cytoplasm (p. 10). They are produced in the bone marrow (↑).

granulocyte (*n*) = polymorphonuclear leucocyte (↑).

eosinophil (*n*) a polymorphonuclear leucocyte (↑) which can be stained with acid (p. 15) dyes such as eosin. Their numbers are normally quite low in the blood (↑) but increase in number if the body becomes infected with parasitic (p. 92) or allergic (p. 234) disease.

basophil (*n*) a polymorphonuclear leucocyte (↑) which can be stained with basic (p. 15) dyes. Their numbers are normally very low in the blood (↑) but they are able to engulf bacteria (p. 42).

neutrophil (*n*) the commonest type of leucocytes (↑) which are able to migrate out of the blood (↑) stream into the tissues (p. 83) of the body to engulf bacteria (p. 42) wherever they invade. On their death, they give rise to pus.

lymphocyte (*n*) a white blood cell (↑) which is produced in the lymphatic system (p. 128) and is important in defending the body against disease. It has a large nucleus (p. 13) and clear cytoplasm (p. 10).

monocyte (*n*) the largest type of white blood cell (↑) and is produced in the lymphatic system (p. 128). It has a spherical nucleus (p. 13) and clear cytoplasm (p. 10). It actively engulfs and devours any invading foreign bodies such as bacteria (p. 42).

nervous tissue tissue (p. 83) containing the nerve cells (p. 149), which are specialized for the transmission of nervous impulses (p. 150), together with the supporting connective tissue (p. 88).

nutrition (*n*) the means by which an organism provides its energy by using nutrients (↓).

nutrient (*n*) any material which is taken in by a living organism and which enables it to grow and remain healthy, replace lost or damaged tissue (p. 83), and provide energy for these and other functions.

holophytic (*adj*) of nutrition (↑), such as that of plants, in which simple inorganic compounds (p.15) can be taken in and built up into complex organic compounds (p. 15) using the energy of light, either to provide energy for metabolism (p. 26) or growth or to make living protoplasm (p. 10).

chemosynthetic (*adj*) of nutrition (↑) in which energy is obtained by a simple inorganic (p. 15) chemical reaction such as the oxidation (p. 32) of ammonia to a nitrite by a bacterium (p. 42).

autotrophic (*adj*) of nutrition (↑) in which simple inorganic compounds (p. 15) are taken in and built up into complex organic compounds (p. 15).

heterotrophic (*adj*) of nutrition (↑), such as that in animals and some fungi (p. 46), in which the organic compounds (p. 15) can only be made from other complex organic compounds which have to be first taken into the body.

saprozoic (*adj*) of nutrition (↑) in which the organism takes in organic compounds (p. 15) only in solution (p. 118) rather than in solid form.

holozoic (*adj*) of nutrition (↑), as found in animals, in which complex organic compounds (p. 15) are broken down into simpler substances which are then used to make body structures or oxidized (p. 32) to supply the organism's energy needs.

saprophytic (*adj*) of nutrition (↑) in which the organism obtains complex organic compounds (p. 15) in solution (p. 118) from dead and/or decaying plant or animal material.

parasitic (*adj*) of nutrition (↑) in which the organism derives its food directly from another living organism at the expense of the host (p. 111) but without necessarily killing it.

types of nutrition

holophytic/autotrophic
green plant

heterotrophic/saprophytic
ink cap fungus

heterotrophic/parasitic
pathogenic bacterium

heterotrophic/holozoic
bird

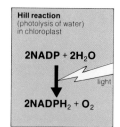

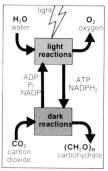

the links between the light reactions and dark reactions of photosynthesis

macronutrient (*n*) any nutrient (↑) which is required by an organism in substantial amounts. *See* p. 240.

major elements = macronutrients (↑).

micronutrient (*n*) a nutrient (↑) which is required in only minute or trace amounts. *See* p. 241.

chlorosis (*n*) the yellowing of the leaves of green plants caused by the loss of chlorophyll (p. 12).

active mineral uptake the uptake and transport through a plant, across a cell membrane (p. 14), of mineral ions from regions of low concentration to regions of high concentration. This process requires energy both to take up minerals and to retain them.

passive mineral uptake the uptake and transport of mineral ions through a plant, usually across a cell membrane (p. 14), from regions of high concentration to regions of low concentration by diffusion (p. 119) without using energy.

photosynthesis (*n*) the process that takes place in green plants in which organic compounds (p. 15) are made from inorganic compounds (p. 15) using the energy of light. It takes place in two main stages: in the light-dependent or photochemical stage, light is absorbed by chlorophyll (p. 12) in the chloroplasts (p. 12), located mainly on the leaves of plants, and used to produce ATP (p. 33) and to supply hydrogen atoms by oxidizing (p. 32) water. These are then used in the reduction (p. 32) of carbon dioxide. In the dark or chemical stage, carbon dioxide is reduced and carbohydrates (p. 17) are made. Photosynthesis will only take place at suitable temperatures and in the presence of chlorophyll, carbon dioxide, water, and light.

limiting factor one of a number of factors which controls the rate at which a chemical reaction, such as photosynthesis (↑), takes place. The rate is limited by that factor which is closest to its minimum or smallest value.

photosynthetic pigment one of the pigments (p. 126) which make up chlorophyll (p. 12) and absorb (p. 81) light. The following substances are photosynthetic pigments: chlorophyll a; chlorophyll b; carotene; and xanthophyll.

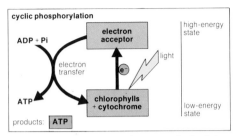

cyclic phosphorylation

products: **ATP**

cyclic photophosphorylation a step in the light dependent stage of photosynthesis (p. 93) in which light is involved in the formation of ATP (p. 33) from ADP (p. 33) by the addition of phosphate.

non-cyclic photophosphorylation a step in the light-dependent stage of photosynthesis (p. 93) in which light is involved in the formation of ATP (p. 33) from ADP (p. 33) by the addition of phosphate and in which the water is split to provide hydrogen ions.

absorption spectrum a diagrammatic representation of the way in which a substance, such as chlorophyll (p. 12), absorbs (p. 81) radiation of different wavelengths by different amounts. Chlorophyll absorbs blue and red light readily so that it appears green.

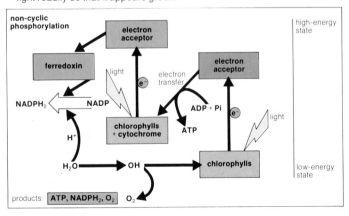

non-cyclic phosphorylation

products: **ATP, NADPH₂, O₂**

action spectra and absorption spectra in photosynthesis

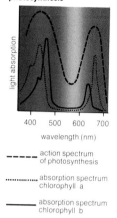

wavelength (nm)

- - - - - action spectrum
of photosynthesis

.... absorption spectrum
chlorophyll a

———— absorption spectrum
chlorophyll b

action spectrum a diagrammatic representation of the way in which radiation of different wavelengths affects a process, such as photosynthesis (p. 93). In this case, it shows that red and blue light are the most effective in the action of photosynthesis.

photosystem I one of the two systems of pigments (p. 126) which each contains chlorophyll a (p. 12), accessory pigments, and electron carriers (p. 31) and which are involved in electron transfer reactions coupled with phosphorylation (↑). Also known as **pigment system I** or **PSI**.

photosystem II *see* photosystem I (↑). Also known as **pigment system II** or **PSII**.

ferredoxin (*n*) any of a number of red-brown (iron-containing) pigments (p. 126) which function as electron carriers (p. 31) in photosynthesis (p. 93).

plastoquinone (*n*) an electron carrier (p. 31) used in photosynthesis (p. 93).

C3 plant a plant in which PGA (p. 97) containing three carbon atoms is produced in the early stage of photosynthesis (p. 93). Photosynthesis in these plants is less efficient than in C4 plants (↓).

C4 plant a plant in which a dicarboxylic acid containing four carbon atoms is produced in the early stage of photosynthesis (p. 93). The method of fixing carbon dioxide has evolved from that of C3 plants (↑) and operates more efficiently.

C_4 pathway of CO_2 fixation

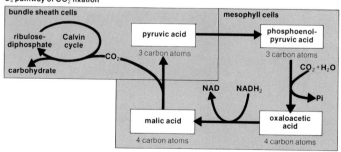

Calvin cycle the steps in the dark stage of
photosynthesis (p. 93) in which carbon dioxide
is reduced (p. 32) using the hydrogen produced
in the light-dependent stage and synthesized
into carbohydrates (p. 17) using the energy of
ATP (p. 33) also formed during the
light-dependent stage.

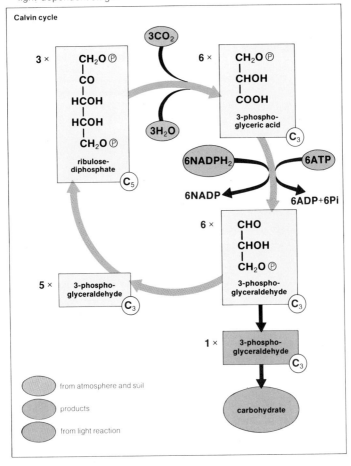

Calvin cycle

compensation point

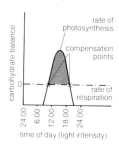

rate of photosynthesis

compensation points

rate of respiration

carbohydrate balance

24.00 6.00 12.00 18.00 24.00

time of day (light intensity)

photorespiration

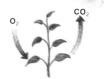

O_2

CO_2

high O_2 concentration in plant tissue

low CO_2 concentration in plant tissue

ribulose diphosphate RUDP. A pentose (p. 17) with which carbon dioxide is combined at the beginning of the Calvin cycle (↑).

phosphoglyceric acid PGA. A complex organic acid (p. 15) which is formed as the result of the combination of carbon dioxide with RUDP (↑) in the fixing of carbon dioxide at the beginning of the Calvin cycle (↑).

phosphoglyceraldehyde (n) a compound formed as the result of the reduction (p. 32) of PGA (↑) during the Calvin cycle (↑). This is then synthesized into starch (p. 18) which is the most important product of photosynthesis (p. 93). Also known as **triose phosphate**.

phosphoenol pyruvic acid PEP. An organic compound (p. 15) which is used by C4 plants (p. 96) in the fixation of carbon dioxide instead of RUDP (↑). Using this compound, carbon dioxide can be stored in chemical form and used later. This is very useful in areas, e.g. the tropics, where carbon dioxide may be in short supply.

compensation point the point at which the intensity of light is such that the amount of carbon dioxide produced by respiration (p. 112) and photorespiration (↓) exactly balances the amount consumed by photosynthesis (p. 93).

photorespiration (n) a light-dependent process in which carbon dioxide is produced and oxygen used up, wasting carbon and energy.

animal nutrition heterotrophic nutrition (p. 92) in which carbohydrates (p. 17) and fats are needed for structural materials and for energy, amino acids (p. 21) are needed to supply nitrogen, and to stimulate growth etc, minerals are required to ensure that the body functions healthily, and vitamins (p. 25) are required to promote and maintain growth.

joule (n) the work done when the point of application of a force of 1 newton is displaced through a distance of 1 metre in the direction of the force. 1 calorie (↓) is equivalent to 4.18 joules. The joule can be used as a measure of the energy value of nutrients (p. 92).

kilojoule (n) = 1000 joules (↑).

calorie (n) see joule (↑).

gut or alimentary canal

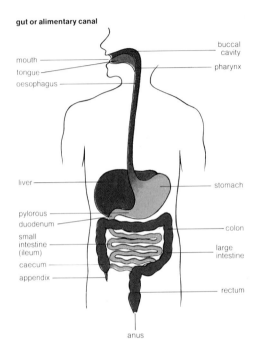

mouth
tongue
oesophagus

buccal cavity
pharynx

liver

stomach

pylorous
duodenum
small intestine (ileum)
caecum
appendix

colon
large intestine
rectum

anus

gut (*n*) a tube, the gastro-intestinal tract, usually leading from the mouth to the anus (p. 103), in animals, which in humans may be as much as 9 metres long and in which food is conveyed, digested (↓) and absorbed (p. 81).

alimentary canal = gut (↑).

ingestion (*n*) the process of taking nutrients (p. 92) into the body for digestion (↓).

digestion (*n*) the breakdown of complex organic compounds (p. 15) or nutrients (p. 92) into simpler, soluble materials which can then be used in the metabolism (p. 26) of the animal.

egestion (*n*) the process of eliminating or discharging food or waste products from the body.

faeces (*n.pl.*) the substances that remain after digestion (↑) and absorption (p. 81) of food in the alimentary canal (↑).

defaecation (*n*) the process of egesting (↑) unwanted food from the body. Material which is defaecated has not taken part in the metabolism (p. 26) of the organism and is therefore not an example of excretion (p. 134).

assimilation (*n*) after digestion (↑), the process of taking into the cells the simple, soluble organic compounds (p. 15) which can then be converted into the complex organic compounds from which the organism is made. **assimilate** (*v*).

buccal cavity the mouth cavity. In mammals (p. 80), that part of the alimentary canal (↑) into which the mouth opens and in which food particles are masticated (p. 104) before they are swallowed.

mucus (*n*) any slimy fluid produced by the mucous membranes (p. 14) of animals and used for protection and lubrication.

saliva (*n*) a fluid secreted into the buccal cavity (↑) by the salivary gland (p. 87) in response to the presence of food. It consists mainly of mucus (↑) and lubricates the food before swallowing. It contains an enzyme (p. 28) in some animals to aid the digestion (↑) of starch (p. 18).

pharynx (*n*) the part of the alimentary canal (↑) between the buccal cavity (↑) and the oesophagus (↓) into which food that has been masticated (p. 104) is pushed by the tongue. The pharynx then contracts by muscular (p. 143) action to force the food into the oesophagus. The gill slits (p. 113) open into the pharynx in fish.

oesophagus (*n*) the part of the alimentary canal (↑) between the pharynx (↑) and the stomach (p.100). It is lined with a folded mucous membrane (p.14) and has layers of smooth muscle fibres (p.144) which contract to force food into the stomach.

epiglottis (*n*) a flap which closes the trachea (p. 175) during swallowing so that food passes into the oesophagus (↑) and not into the trachea.

bolus (*n*) a rounded mass consisting of masticated (p. 104) food particles and saliva (↑) into which food is formed in the buccal cavity (↑) before swallowing.

stomach (*n*) the part of the alimentary canal (p. 98) between the oesophagus (p. 99) and the duodenum (↓) into which food is passed and can be stored in quite large amounts for long periods so that it is not necessary for the animal to be eating continuously. Food is mixed with gastric juices and, although little of it is absorbed (p. 81), materials, such as minerals or vitamins (p. 25), may be taken into the blood (p. 90) stream. The stomach is muscular (p. 143) and is lined with a mucous membrane (p. 14).

peristalsis (*n*) the waves of rhythmical contractions which take place in the alimentary canal (p. 98) by muscular (p. 143) action and which force food through the canal.

chief cell one of a number of cells found in the gastric glands (p. 87) which secrete (p. 106) the enzymes (p. 28) pepsin (p. 107) and rennin (p. 106) which digest (p. 98) proteins (p. 21) and milk protein (in young mammals (p. 80)) respectively. Also known as **peptic cell**.

fundis gland a gland (p. 87) in the stomach (↑) which secretes (p. 106) mucus (p. 99) to protect and lubricate the wall of the stomach.

peristalsis

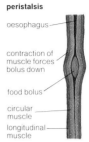

oesophagus

contraction of muscle forces bolus down

food bolus

circular muscle

longitudinal muscle

wave of contraction passing down oesophagus

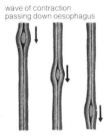

section of stomach wall

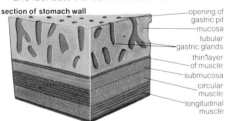

opening of gastric pit
mucosa
tubular gastric glands
thin layer of muscle
submucosa
circular muscle
longitudinal muscle

detail of gastric gland

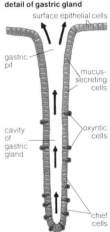

surface epithelial cells

gastric pit

mucus-secreting cells

cavity of gastric gland

oxyntic cells

chief cells

oxyntic cell one of a number of cells found in the gastric glands (p. 87) which secrete (p. 106) hydrochloric acid (HCl) which in turn kills harmful bacteria (p. 42), makes available calcium and iron salts and provides a suitably low pH (p. 15) for the formation of pepsin (p. 107).

chyme (*n*) a partially broken-down, semi-fluid mixture of food particles and gastric juices which is then released in small quantities into the duodenum (↓).

section through lining of duodenum

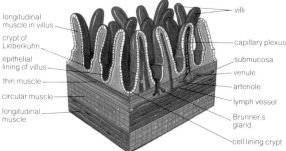

- villi
- capillary plexus
- submucosa
- venule
- arteriole
- lymph vessel
- Brunner's gland
- cell lining crypt

- longitudinal muscle in villus
- crypt of Lieberkuhn
- epithelial lining of villus
- thin muscle
- circular muscle
- longitudinal muscle

duodenum (*n*) the part of the alimentary canal (p. 98) between the stomach (↑) and the ileum (p. 102) which forms the first part of the small intestine (p. 102) into which the chyme (↑) passes from the stomach. Digestion (p. 98) continues in the duodenum with the aid of intestinal juices (p. 102) and, in addition, it receives secretions (p. 106) from the pancreas (p. 102) and the liver (p. 103). Its lining is covered with villi (p. 103) and the glands (p. 87) that secrete the intestinal juices.

chyle (*n*) lymph (p. 128) containing the results of digestion (p. 98). The liquid looks milky because it contains emulsified (p. 26) fats and oils.

bile (*n*) a secretion (p. 106) from the liver (p. 103) which contains some waste material from the liver and bile salts which emulsify fats, increase the activity of certain enzymes (p. 28), aid in the absorption (p. 81) of some vitamins (p. 25) and is rich in sodium bicarbonate which neutralizes stomach (↑) acid.

bile duct the tube through which bile (↑) is passed from the liver (p. 103) to the duodenum (↑).

gall bladder a sac-like bladder extending from the bile duct (↑) and situated within or near the liver (p. 103). It functions as a store for bile (↑) when it is not required for digestive (p. 98) purposes and then, by muscular (p. 143) contractions, empties into the duodenum (↑) through the bile duct.

pancreatic juice a solution (p. 118) in water of alkaline salts, which neutralize the acid (p. 15) from the stomach (p. 100), and enzymes (p. 28) to aid in digestion (p. 98).

pancreas (*n*) a gland (p. 87) which is connected to the duodenum (p. 101) by a duct and which produces pancreatic juice (↑) and insulin (↓).

islets of Langerhans cells contained within the pancreas (↑) which produce insulin (↓).

insulin a hormone (p. 130) which controls the sugar level in the blood (p. 90) and, if it is deficient, the sugar level rises while, if it is in excess, the level falls leading to a coma.

jejunum (*n*) the part of the small intestine (↓) between the duodenum (p. 101) and the ileum (↓).

small intestine the narrow tube which forms part of the alimentary canal (p. 98) between the stomach (p. 100) and the colon (↓). It is the main region of digestion (p. 98) and absorption (p. 81) and includes the duodenum (p. 101).

ileum (*n*) the longest, usually coiled part of the small intestine (↑) between the jejunum (↑) and the colon (↓). It is muscular (p. 143) and causes food particles to pass along it by peristalsis (p. 100). Its lining is folded and covered with large numbers of villi (↓) which increase the surface area for absorption (p. 81).

intestinal juice a secretion (p. 106) produced in the glands (p. 87) of the intestine (↑) which contains a mixture of enzymes (p. 28) of digestion (p. 98), such as amylase (p. 106) and sucrase (p. 107). Also known as **succus entericus**.

Brunner's glands the deep-lying glands (p. 87) in the walls of the duodenum (p. 101) which secrete (p. 106) intestinal juices.

crypts of Lieberkuhn the glands (p. 87) found within the walls of the small intestine (↑) which secrete (p. 106) the intestinal juices.

appendix (*n*) in humans, a blind-ended tube at the end of the caecum (↓).

caecum (*n*) a blind-ended branch of the gut (p. 98) at the junction of the small and large intestines (↑). It is very large and important in the digestions (p. 98) of some mammals (p. 80), not humans.

**section across
small intestine (ileum)**

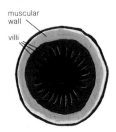

muscular
wall

villi

villus

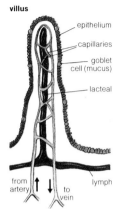

epithelium

capillaries

goblet cell (mucus)

lacteal

from artery

to vein

lymph

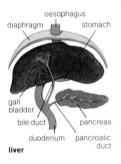

oesophagus

diaphragm

stomach

gall bladder

bile duct

pancreas

duodenum

pancreatic duct

liver

colon (*n*) the first part of the large intestine. It secretes (p. 106) mucus (p. 99) and contains the remains of food materials which cannot be digested (p. 98) as well as the digestive juices. From this material, water and vitamins (p. 25) are reabsorbed into the blood (p. 90) leaving the faeces (p. 99) which are moved on to the rectum (↓).

rectum (*n*) the part of the large intestine in which the faeces (p. 99) are stored before release through the anus (↓).

anus (*n*) the posterior opening to the alimentary canal (p. 98) through which the faeces (p. 99) may pass at intervals and which is closed by a ring of muscles (p. 143) called the anal sphincter (p. 127).

villi (*n.pl.*) the rod-like projections which cover the lining of the small intestine (↑) to increase the surface area for absorption (p. 81). **villus** (*sing.*).

liver (*n*) a gland (p. 87) which lies close to the stomach (p. 100) and is connected with the small intestine (↑) by the bile duct (p. 101) through which it secretes (p. 106) bile (p. 101) for digestive (p. 98) purposes. The liver also removes damaged red cells (p. 91) from the blood (p. 90), stores iron, synthesizes vitamin (p. 25) A, and stores vitamins A, D, and B, synthesizes blood proteins (p. 21), removes blood poisons, and synthesizes agents which help blood clot (p. 129), breaks down alcohol, stores excess carbohydrates (p. 17) and metabolizes (p. 26) fats.

hepatic portal vein one of a system of veins (p. 127) which carry blood (p. 90) rich in absorbed (p. 81) food materials, such as glucose (p. 17), direct from the intestine (↑) to the liver (↑).

liver cell one of the cells making up the liver (↑). Each cell is in direct contact with the blood (p. 90) so that material diffuses between blood and liver very rapidly. Liver cells are cube shaped with granular cytoplasm (p. 10).

reticulo-endothelial system a system of macrophage (p. 88) cells which is present in the liver (↑) and other parts of the body and which is in contact with the blood (p. 90) and other fluids. These macrophage cells are able to engulf foreign bodies and thus protect the body from infection, damage and disease.

tooth (n) one of a number of hard, resistant structures found growing on the jaws (↓) of vertebrate (p. 74) animals and which are used to break down food materials mechanically. They may be specialized for different functions among different animals and even within the same animal. **teeth** (pl.).

dentition (n) the kind, arrangement and number of the teeth (↑) of an animal.

heterodont dentition the condition in which the teeth (↑) of an animal, typically mammals (p. 80), are differentiated into different forms, such as molars (↓) and canines (↓), to perform different functions, such as grinding up the food or killing prey.

homodont dentition the condition in which the teeth (↑) of an animal are all identical.

mastication (n) the process which takes place in the buccal cavity (p. 99) whereby food is mechanically broken down by the action of teeth (↑), tongue, and cheeks into a bolus (p. 99) for swallowing.

dental formula a formula which indicates by letters and numbers the types and numbers of teeth (↑) in the upper and lower jaws (↓) of a mammal (p. 80). For example, the dental formula of a human would be
i2/2, c1/1, p2/2, m3/3
which indicates that both the upper and lower jaws have two incisors (↓), one canine (↓), two premolars (↓), and three molars (↓) on each side of the jaw.

incisor (n) one of the chisel-shaped teeth (↑), very prominent in rodents, that occur at the very front of a mammal's (p. 80) jaw (↓) and which have a single root and a sharp edge with which to sever portions of the food.

canine (n) one of the 'dog teeth' or pointed conical teeth (↑) which occur between the incisors (↑) and the premolars (↓) and which are used to kill prey in carnivorous (p. 109) animals such as dogs and cats.

carnassial (n) one of the large flesh-cutting teeth (↑) found in terrestrial (p. 219) carnivores (p. 109).

heterodont dentition
e.g. carnivore (a dog)

incisors
molars
upper jaw (maxilla)
canines
carnassials
premolars
lower jaw (mandible)

incisor

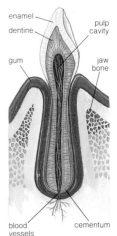

enamel
pulp cavity
dentine
gum
jaw bone
blood vessels and nerve fibres
cementum

molar

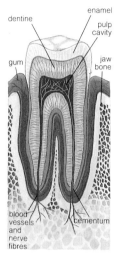

dentine

gum

enamel

pulp
cavity

jaw
bone

blood
vessels
and
nerve
fibres

cementum

premolar (*n*) one of the crushing and grinding
cheek teeth (↑) that occur between the canines
(↑) and molars (↓) in mammals (p. 80). They are
usually ridged and furrowed with more than
one root. They are represented in the first 'milk'
or deciduous (p. 59) dentition (↑).

molar (*n*) one of the large crushing and grinding
cheek teeth (↑) that occur at the back of the
mouth of a mammal (p. 80). They are ridged
and furrowed with more than one root. They are
not represented in the first 'milk' or deciduous
(p. 59) dentition (↑) and, in humans, there are
four which do not erupt until later in life and are
referred to as 'wisdom teeth'.

enamel (*n*) the hard outer layer of the tooth (↑) of
a vertebrate (p. 74). It is composed mainly of
crystals of the salts, carbonates, phosphates,
and fluorides of calcium held together by small
amounts of an organic compound (p. 15).

dentine (*n*) the hard substance that makes up the
bulk of the tooth (↑) in mammals (p. 80). It is
similar to bone but has a higher mineral content
and no cells.

cementum (*n*) the hard substance that covers
the root of the tooth (↑) in mammals (p. 80). It is
similar to bone but has a higher mineral content
and lacks Haversian canals (p. 89).

pulp cavity the substance within the centre of a
tooth (↑) which contains the blood vessels
(p. 127) and nerves (p. 149) which supply the
tooth, together with connective tissue (p. 88). It
connects to the tissue (p. 83) into which the
tooth is fixed.

gum (*n*) the tissue (p. 83) that surrounds and
supports the roots of the teeth (↑) and covers
the jaw (↓) bones. It contains nerves (p. 149)
and many blood capillaries (p. 127) giving it the
characteristic pink colour when it is healthy.
Also known as **gingiva**.

jaw (*n*) one of the bones in which the teeth (↑) are
set. The jaw movements as well as the dentition
(↑) of different animals are specialized for
different actions, for example, tearing,
snatching and chewing movements, such as
crushing and grinding.

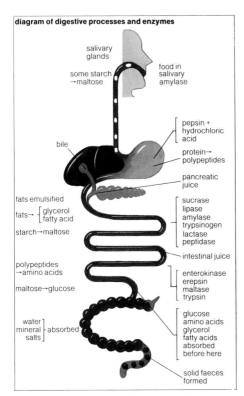

diagram of digestive processes and enzymes

salivary glands

some starch →maltose

food in salivary amylase

bile

pepsin + hydrochloric acid

protein→ polypeptides

pancreatic juice

fats emulsified

fats→ [glycerol / fatty acid]

starch→maltose

sucrase
lipase
amylase
trypsinogen
lactase
peptidase

intestinal juice

polypeptides →amino acids

maltose→glucose

enterokinase
erepsin
maltase
trypsin

water
mineral | absorbed
salts

glucose
amino acids
glycerol
fatty acids
absorbed
before here

solid faeces formed

secretion (*n*) a material with a special function in an organism which is made within a cell and passed out of the cell to perform its function. **secrete** (*v*).

amylase (*n*) any of a number of enzymes (p. 28) which catalyze (p. 28) the hydrolysis (p. 16) of carbohydrates (p. 17), such as starch (p. 18), into simple sugars. It is secreted (↑) as saliva (p. 99), and in the pancreas (p. 102) and the small intestine (p. 102).

rennin (*n*) an enzyme (p. 28) which coagulates (p. 128) milk. It is secreted (↑) by the gastric glands (p. 87) of the stomach (p. 100).

maltase (*n*) an enzyme (p. 28) which catalyzes
(p. 28) the hydrolysis (p. 16) of maltose (p. 18)
into two molecules of glucose (p. 17). It is
secreted (↑) by the small intestine (p. 102).

lactase (*n*) an enzyme (p. 28) which catalyzes (p. 28)
the hydrolysis (p. 16) of the disaccharide (p. 18)
lactose (p. 18) into glucose (p. 17) and galactose
(p.18). It is secreted (↑) by the small intestine (p.102).

sucrase (*n*) an enzyme (p. 28) which catalyzes (p. 28)
the hydrolysis (p. 16) of sucrose (p. 18) into glucose
(p. 17) and fructose (p. 17). It is secreted (↑) by the
small intestine (p. 102). Also known as **invertase**.

erepsin (*n*) a mixture of enzymes (p. 28) which
catalyzes (p. 28) the breakdown of proteins
(p. 21) into amino acids (p. 21). It is secreted (↑)
by the small intestine (p. 102).

lipase (*n*) an enzyme (p. 28) which catalyzes
(p. 28) the hydrolysis (p. 16) of fats into fatty
acids (p. 20) and glycerol (p. 20). It is secreted
(↑) by the pancreas (p. 102).

enterokinase (*n*) an enzyme (p. 28) which
catalyzes (p. 28) the conversion of trypsinogen
(p. 108) into trypsin (↓). It is secreted (↑) by the
small intestine (p. 102).

chymotrypsin (*n*) an enzyme (p. 28) which
catalyzes (p. 28) the conversion of proteins
(p. 21) into amino acids (p. 21). It is secreted (↑)
by the pancreas (p. 102).

pepsin (*n*) an enzyme (p. 28) which catalyzes
(p.28) the hydrolysis (p. 16) of proteins (p. 21)
into polypeptides (p. 21) in acid solution (p.118).
It is secreted (↑) by the stomach (p. 100) as
pepsinogen (↓).

pepsinogen (*n*) the inactive form of pepsin (↑)
which is secreted (↑) by the stomach (p. 100)
and activated by hydrochloric acid (HCl).

gastrin (*n*) a hormone (p. 130) which stimulates
the secretion (↑) of hydrochloric acid (HCl)
and pepsin (↑) in the stomach (p. 100). It is
activated by the presence of food materials.

trypsin (*n*) an enzyme (p. 28) which catalyzes
(p. 28) the hydrolysis (p. 16) of proteins (p. 21)
into polypeptides (p.21) and amino acids (p.21).
It is secreted (↑) by the pancreas (p. 102) as
trypsinogen (p. 108).

trypsinogen (*n*) the inactive form of trypsin (p. 107)
which is secreted (p. 106) by the pancreas (p. 102)
and converted into trypsin by enterokinase (p. 107).

peptidase (*n*) an enzyme (p. 28) which catalyzes
(p. 28) the hydrolysis (p. 16) of polypeptides
(p. 21) into amino acids (p. 21) by breaking
down the peptide bonds (p. 21). It is secreted
(p. 106) by the small intestine (p. 102).

nucleotidase (*n*) an enzyme (p. 28) which catalyzes
(p. 28) the hydrolysis (p. 16) of a nucleotide (p. 22)
into its component nitrogen bases (p. 22), pentose
(p. 17) and phosphoric acid (p. 22). It is secreted
(p. 106) by the small intestine (p. 102).

secretin (*n*) a hormone (p. 130) which stimulates
the secretion (p. 106) of bile (p. 101) from the
liver (p. 103) and digestive (p. 98) juices from
the pancreas (p. 102). It is secreted by the
duodenum (p. 101).

pancreozymin (*n*) a hormone (p. 130) which
stimulates the release of digestive (p. 98) juices
from the pancreas (p. 102). It is ꞏecreted
(p. 106) by the duodenum (p. 101).

microphagous (*adj*) of an animal which feeds on
food particles that are tiny compared with the
size of the animal so that it must feed
continuously to receive enough nutrients (p. 92).

filter feeder a microphagous (↑) feeder which
lives in water and filters suspended food
particles from the water.

deposit feeder a microphagous (↑) feeder which
feeds on particles that have been deposited on
and perhaps mixed with the base layer of the
environment (p. 218) in which the animal lives.

fluid feeder a microphagous (↑) feeder which
feeds by ingesting (p. 98) fluids containing
nutrients (p. 92) from living or recently dead
animals or plants.

pseudopodial feeder a microphagous (↑) feeder
in which cells develop temporary, finger-like
projections, pseudopodia (p. 44), to engulf food
particles.

mucous feeder a microphagous (↑) feeder in
which food particles are trapped in mucus
(p. 99) secreted (p. 106) by the organism and
moved by ciliary (p. 12) action to the mouth.

types of feeding
microphagous feeder

e.g. Right whale

fluid feeder e.g. mosquito

pseudopodial feeder
e.g. *Amoeba*

mucous feeder
e.g. lancelet
mouth with
tentacles

types of feeding

setous feeder
e.g. *Daphnia*

macrophagous feeders

omnivore
e.g. American oppossum

carnivore
e.g. tiger

ruminant herbivore
e.g. gazelle

gut of a bird

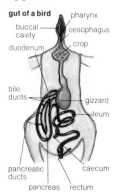

pharynx
buccal cavity
oesophagus
crop
duodenum
bile ducts
gizzard
ileum
pancreatic ducts
caecum
pancreas
rectum

setous feeder a microphagous (↑) feeder in which the food particles are trapped by setae (p. 65) and then moved towards the mouth by beating cilia (p. 12).

macrophagous (*adj*) of an animal which feeds on relatively large food particles and usually, therefore, does not need to feed continuously.

coprophagous (*adj*) of an animal, such as some rodents, which feed on faeces (p. 99) thus improving the digestion (p. 98) of cellulose (p. 19) on second passage.

omnivore (*n*) an animal which feeds by eating a mixed diet of animal and plant food material.

carnivore (*n*) an animal which feeds by eating a diet that consists mainly of animal material. Carnivores may have powerful claws and dentition (p. 104) adapted to tearing flesh.

herbivore (*n*) an animal which feeds by eating a diet that consists mainly of plant material. Herbivores may have dentition (p. 104) and digestion (p. 98) specially adapted to deal with tough plant materials.

ruminant (*n*) one of the group of herbivores (↑) which belong to the order Artiodactyla and in which the stomach (p. 100) is complex and includes a rumen. Food is eaten but not chewed initially, and it is passed to the rumen where it is partly digested (p. 98) and then regurgitated for further chewing before swallowing and passing into the reticulum.

gizzard (*n*) part of the alimentary canal (p. 98) of certain animals. It has a very tough lining surrounded by powerful muscles (p. 143) in which food particles are broken down by grinding action against grit or stones in the gizzard lining or against spines or 'teeth' in the gizzard itself.

crop (*n*) the part of the alimentary canal (p. 98) in animals, such as birds, which either forms part of the gut (p. 98) or the oesophagus (p. 99), and in which food is stored temporarily and partly digested (p. 98).

mandible (*n*) (1) the lower jaw (p. 105) of a vertebrate (p. 74); (2) either of the pair of feeding mouth parts of certain invertebrates (p. 75).

carnivorous plant any plant which supplements its supply of nitrates by capturing, using a variety of means, small animals, such as insects (p. 69) and digesting (p. 98) them with enzymes (p. 28) secreted (p. 106) externally.

parasitism (*n*) an association (p. 227) in which the individuals of one species (p. 40), the parasites, live permanently or temporarily on individuals of another species, the host (↓), deriving benefit and/or nutrients (p. 92) and causing harm or even death to the host.

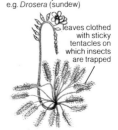

carnivorous plant
e.g. *Drosera* (sundew)

leaves clothed with sticky tentacles on which insects are trapped

parasitism e.g. infestation by hookworms

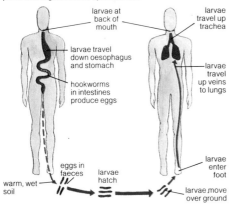

larvae at back of mouth

larvae travel up trachea

larvae travel down oesophagus and stomach

larvae travel up veins to lungs

hookworms in intestines produce eggs

eggs in faeces

larvae hatch

warm, wet soil

larvae enter foot

larvae move over ground

parasite (*n*) *see* parasitism (↑).

endoparasite (*n*) a parasite (↑) which lives within the body of the host (↓) itself, for example, the tapeworm, living within the gut (p. 98) of a vertebrate (p. 74).

ectoparasite (*n*) a parasite (↑) which lives on the surface of the host (↓) and is usually adapted for clinging on to the host on which it often feeds by fluid feeding (p. 108). Ectoparasites usually have special organs for attachment to the host.

intercellular parasite an endoparasite (↑) which lives within the material between the cells of the host (↓).

intercellular (*adj*) between cells.

ectoparasite e.g. dodder (*Cuscuta*) on a bean. The dodder climbs around the host stem and taps into its vascular system via haustoria

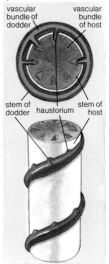

vascular bundle of dodder

vascular bundle of host

stem of dodder

haustorium

stem of host

endoparasite e.g. fungal pathogen. Fungal hyphae weave between cells and tap cells for nutrients with haustoria

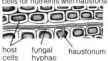

host cells

fungal hyphae

haustorium

intracellular parasite an endoparasite (↑)which lives within the cells of the host (↓).

host (*n*) the species (p. 40) of organism in an association (p. 227) within which or on which a parasite (↑) lives and reaches sexual maturity, and which suffers harm or even death as a result.

secondary host a host (↑) on which or within which the young or resting stage of a parasite (↑) may live temporarily. The parasite does not reach sexual maturity on the secondary host. Also known as **intermediate host**.

transmission (*n*) the process by which a substance or an organism is transported from one place to another, e.g. a parasite (↑) is transmitted from one host (↑) to another sometimes via a secondary host (↑) and it may involve considerable risk to the parasite. **transmit** (*v*).

vector (*n*) a secondary host (↑) which is actively involved in the transmission (↑) of a parasite (↑) from one host (↑) to another or an organism which passes infectious disease from one individual to another without necessarily being affected by the disease itself. For example, the blood-sucking mosquito which transmits a malaria-causing blood parasite from one individual on which it feeds to another is a vector.

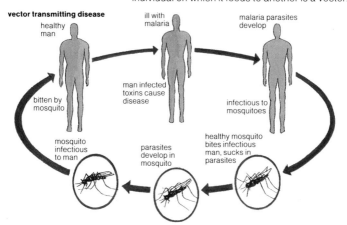

vector transmitting disease

healthy man

ill with malaria

malaria parasites develop

man infected toxins cause disease

bitten by mosquito

infectious to mosquitoes

mosquito infectious to man

parasites develop in mosquito

healthy mosquito bites infectious man, sucks in parasites

respiration (*n*) the process whereby energy is produced in a plant or animal by chemical reactions. In most organisms, this is achieved by taking in oxygen from the environment (p. 218) and, after transportation to the cells, its reaction with food molecules releases carbon dioxide, water and energy that is trapped in ATP (p. 33) – cellular respiration (p. 30).

respiratory quotient RQ. The ratio of the volume of carbon dioxide produced by an organism to the volume of oxygen used up during the same period of respiration.

$$RQ = \frac{\text{carbon dioxide produced}}{\text{oxygen used up}}$$

breathing (*n*) the process of actively drawing air or any other gases into the respiratory (↑) organs for gas exchange (↓). **breathe** (*v*).

gas exchange the process which takes place at a respiratory surface (↓) in which a gas, such as oxygen, from the environment (p. 218) diffuses (p. 49) into the organism because the concentration of that gas in the organism is lower than in the environment, and another gas, such as carbon dioxide, is released from the organism into the environment. In plants, respiratory gas exchange is complicated by the gas exchange that takes place as a result of photosynthesis (p. 93).

respiratory surface the surface of an organ, such as a lung (p. 115), across which gas exchange (↑) takes place. It is usually highly folded to increase the surface area, thin, and, in organisms that live on land, damp.

inspiration (*n*) the process of drawing air into or across the respiratory surface (↑) by muscular (p. 143) action. The pressure within the respiratory organ is reduced to below that of the environment (p. 218) so that air flows in. Also known as **inhalation**. **inspire** (*v*).

expiration (*n*) the process of forcing air and waste gases out of the respiratory (↑) organ by muscular (p. 143) action. The pressure within the respiratory organ is increased so that air flows out. Also known as **exhalation**. **expire** (*v*).

air (*n*) the mixture of gases which forms the atmosphere surrounding the Earth. It is composed of approximately 78 per cent nitrogen, 21 per cent oxygen, 0.03 per cent carbon dioxide, and little under 1 per cent of the so-called noble gases including argon, neon, etc. It also includes water vapour.

gill (*n*) one part of the respiratory surface (↑) found in most aquatic animals, such as fish. Gills are projections of the body wall or of the inside of the gut (p. 98) and may be very large and complex in relation to the animal because they are supported by water. They are very thin and well supplied with blood vessels (p. 127) so that gas exchange (↑) between the water and the blood (p. 90) of the animal is usually very efficient.

two main gill types showing water movements

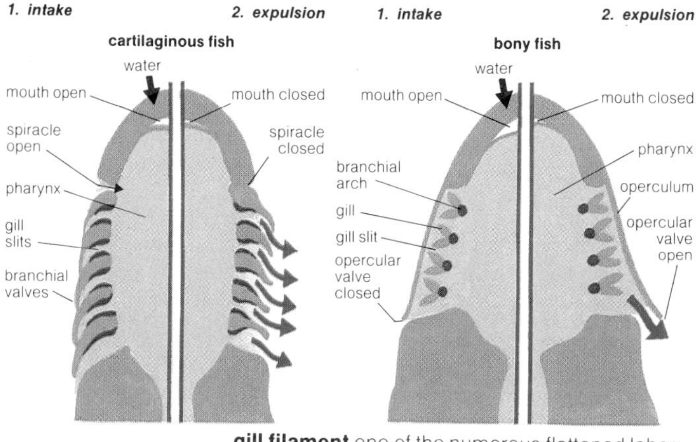

1. intake   *2. expulsion*   *1. intake*   *2. expulsion*

cartilaginous fish

bony fish

water
mouth open
spiracle open
pharynx
gill slits
branchial valves
mouth closed
spiracle closed

water
mouth open
branchial arch
gill
gill slit
opercular valve closed
mouth closed
pharynx
operculum
opercular valve open

gill filament one of the numerous flattened lobes which make up a gill and increase its surface area.

gill slit one of a series of openings through the pharynx (p. 99) in fish and some amphibians (p. 77) leading to the gills (↑).

operculum[a] (*n*) a bony plate that covers the visceral clefts (p. 74) and the gill slits (↑) in bony fish (p. 76). It assists in pumping water over the gills (↑) for gas exchange (↑) by inward and outward movements.

counter current exchange system the system found in the gills (p. 113) of bony fish (p. 76) in which the water that is pumped past the gill filaments (p. 113) flows in the opposite direction to the flow of blood (p. 90) within the gill. Gas exchange (p. 112) takes place continuously along the whole length of the gill because the levels of gases never reach equilibrium.

parallel current exchange system the system of gas exchange (p. 112) found in the gills (p. 113) of cartilaginous fish (p. 76),in which the flow of water and the flow of blood (p. 90) are in the same direction. This system is less efficient than a counter current exchange system (↑) because equilibrium is soon reached.

buccal pump one part of the double pump action which causes oxygen containing water to flow over the gills (p. 113). Muscles (p. 143) in the floor of the buccal cavity (p. 99) cause it to rise and fall as the mouth shuts and opens forcing water over the gills and then drawing in more water through the mouth respectively.

opercular pump one part of the double pump action which causes oxygen-containing water to flow over the gills (p. 113). Muscles cause the operculum (p. 113) to open outwards as water is drawn in through the mouth.

tracheal system in insects (p. 96), a system of gas exchange (p. 112) and transport which is separate from the blood (p. 90) system. Oxygen is carried by air tubes called tracheae (↓) and some of the oxygen in the air diffuses into the body tissues (p. 83). Oxygen is also dissolved in the fluid in the tracheoles (↓).

spiracle (n) the opening to the atmosphere of an insect's (p. 69) trachea(↓).

counter current exchange system

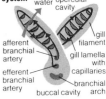

water opercular cavity · gill filament · afferent branchial artery · gill lamella with capillaries · efferent branchial artery · branchial arch · buccal cavity

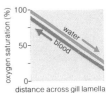

oxygen saturation (%) · water · blood · distance across gill lamella

parallel current exchange system

vertical septum · gill filament · gill lamella with capillaries · afferent branchial artery · efferent branchial artery · branchial arch · water · buccal cavity

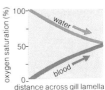

oxygen saturation (%) · water · blood · distance across gill lamella

tracheal system

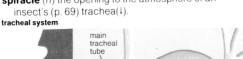
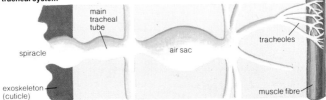

main tracheal tube · spiracle · air sac · tracheoles · exoskeleton (cuticle) · muscle fibre

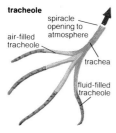

tracheole

spiracle opening to atmosphere

air-filled tracheole

trachea

fluid-filled tracheole

trachea (*n*) (1) in insects (p. 69), one of a number of air tubes which lead from the spiracles (↑) into the body tissues (p. 83); (2) in land-living vertebrates (p. 74), a tube which leads from the throat into the bronchi (p. 116). **tracheae** (*pl.*).

tracheole (*n*) one of the very fine tubes into which the tracheae (↑) of insects (p. 69) branch. They pass into the muscles (p. 143) and organs of an insect's body to allow gas exchange (p. 112) to take place.

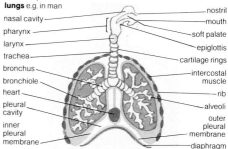

lungs e.g. in man

nasal cavity — pharynx — larynx — trachea — bronchus — bronchiole — heart — pleural cavity — inner pleural membrane —

nostril — mouth — soft palate — epiglottis — cartilage rings — intercostal muscle — rib — alveoli — outer pleural membrane — diaphragm

lungs (*n.pl.*) a pair of thin-walled, elastic sacs present in the thorax (↓) of amphibians (p. 77), reptiles (p. 78), birds and mammals (p. 80) and containing the respiratory surfaces (p. 112).

ventilation (*n*) the process whereby the air contained within the lungs (↑) is exchanged with air from the atmosphere by regular breathing (p. 112) in which muscular (p. 143) movements of the thorax (↓) varies its volume and thus the volume of the lungs. During inspiration (p. 112) the volume of the lungs increases and atmospheric pressure forces air into the lungs. During expiration (p. 112), the muscles relax and the volume of the lungs decreases by virtue of their elasticity so that air is forced out.

thorax (*n*) (1) in arthropods (p. 67), the segments between the head and the abdomen (p. 116); (2) in vertebrates (p. 74), the part of the body which contains the heart (p. 124) and lungs (↑). In mammals (p. 80) it is separated from the abdomen by the diaphragm (p. 116) and protected by the rib cage.

thoracic cavity = thorax (↑), in vertebrates (p. 74).

intercostal muscle a muscle (p. 143) which connects adjacent ribs. When the external intercostal muscles contract, the ribs are moved upwards and outwards, increasing the volume of the thoracic cavity (p. 115) and thus the lungs (p. 115) so that air is forced into the lungs for inspiration (p. 112). When the internal intercostal muscles contract the volume of the thoracic cavity decreases and expiration (p. 112) takes place.

diaphragm (*n*) a sheet of muscular (p. 143) tissue (p. 83) which separates the thoracic cavity (p.115) from the abdomen (↓) in mammals (p. 80).

abdomen (*n*) (1) in arthropods (p. 67), the segments at the back of the body; (2) in vertebrates (p. 74), the part of the body containing the intestines, liver, kidney etc.

pleural cavity the narrow space, filled with fluid, between the two layers of the pleural membrane (↓).

pleural membrane the double membrane which surrounds the lungs (p. 115) and lines the thoracic cavity (p. 115). It secretes (p. 106) fluids to lubricate the two layers as the lungs expand and contract during breathing (p. 112).

larynx (*n*) a structure found at the junction of the trachea (p. 115) and the pharynx (p. 99) which contains the vocal cords (↓). During swallowing it is closed off by the epiglottis (p. 99).

vocal cord (*n*) one of the folds of the lining of the larynx (↑) which produce sound as a current of air passes over them.

bronchus (*n*) one of the two large air tubes into which the trachea (p. 115) divides and which enter the lungs (p. 115).

bronchiole (*n*) one of a number of smaller air tubes into which the bronchi (↑) divide after entering the lungs (p. 115). The bronchioles make up the 'bronchial tree' which ends in air tubes called the *respiratory bronchioles*. These divide into *alveolar ducts* (or terminal bronchioles) which give rise to the alveoli (↓).

alveolus (*n*) a pouch-like air sac which occurs in clusters at the ends of the bronchioles (↑) and which contains the respiratory surfaces (p. 112). A network of capillaries (p. 127) covers the thin, elastic epithelium (p.87). **alveoli** (*pl.*)

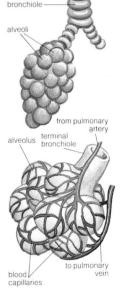

bronchiole and alveoli

bronchiole

alveoli

from pulmonary artery

alveolus

terminal bronchiole

to pulmonary vein

blood capillaries

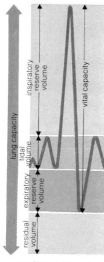

**comparison of different
lung volumes in man**

tidal flow of the system in which inspiration
(p. 112) and expiration (p. 112) take place
through the same air passages so that the air
passes twice over each part of the respiratory
surface (p. 112). This is less efficient than a
system in which there is a constant throughflow
such as that which takes place over the gills
(p. 113) of fish.

tidal volume the volume of air which is inspired
(p. 112) or expired (p. 112) during normal
regular breathing (p. 112). It is considerably
less than the lung capacity (↓).

ventilation rate the rate per minute at which the
total volume of air is expired (p. 112) or inspired
(p. 112).

residual volume the volume of air that always
remains within the alveoli (↑) because the
thorax (p. 115) is unable to collapse completely.
It exchanges oxygen and carbon dioxide with
the tidal air.

vital capacity the total amount of air which can
be inspired (p. 112) and expired (p. 112) during
vigorous activity.

reserve volume the difference in volume
between the total lung capacity (↓) and the vital
capacity (↑).

lung capacity the total volume of air that can be
contained by the lungs when fully inflated.

acclimatization (n) the period of time it takes for
the respiration (p. 112) of an organism to get
used to the reduced partial pressure of oxygen
that may occur at high altitudes, for example,
where the atmospheric pressure is reduced.

respiratory centre the part of the medulla
oblongata (p. 156) which controls the rate of
breathing (p. 112) in response to the levels of
carbon dioxide dissolved in the bloodstream
(p. 90).

oxygen debt the deficit in the amount of oxygen
that is available for respiration (p. 112) during
vigorous activity so that even when the activity
ceases, breathing (p. 112) continues at a high
rate until the oxygen debt is made up. Lactic
acid builds up in the muscles (p. 143) from
lactic acid fermentation (p. 34).

osmosis (*n*) the process by which water passes
through a semipermeable membrane (↓) from
a solution (↓) of low concentration of salts to
one of high concentration thereby diluting it.
Osmosis will continue until the concentrations
of the two solutions are equalized. In living
things osmosis can take place through
membranes (p. 14), e.g. tonoplast (p. 11) or
plasmalemma (p. 14), in either direction. In
plants, the cell walls (p. 8) are elastic so that
they can contain solutions of higher concentration
when osmosis ceases. **osmotic** (*adj*).

osmosis   semipermeable membrane

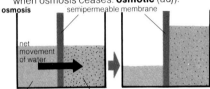

net
movement
of water

pure water   concentrated
or hypotonic   or hypertonic
solution with   solution with
a high osmotic   a low osmotic
pressure   pressure

osmotic potential the tendency of water
molecules to diffuse (↓) through a semipermeable
membrane (↓) from a solution (↓) of low solute
concentration to a solution of high solute
concentration until equilibrium is reached.

semipermeable membrane a membrane (p. 14),
such as a tonoplast (p. 11) or plasmalemma (p. 14),
with microscopic (p. 9) pores (p. 120) through which
small molecules e.g. water will pass but larger
molecules e.g. sucrose (p. 18), or salts will not.

solution (*n*) a liquid (*the solvent*) with substances
(*the solute*) dissolved in it. Substances that will
dissolve are said to be *soluble* and those that
will not, *insoluble*.

isotonic solution (*n*) a solution (↑) in which the
osmotic potential (↑) is the same as that of
another solution so that neither solution either
gains or loses water by osmosis (↑) across a
semipermeable membrane (↑).

hypotonic (*adj*) of one solution (↑) in an osmotic
(↑) system which is more dilute than another.

hypertonic (*adj*) of one solution (↑) in an osmotic
(↑) system which is more concentrated than another.

diffusion (*n*) the process in which molecules move from an area of high concentration to an area of low concentration. Osmosis (↑) is a special type of diffusion restricted to the movement of water molecules.

diffusion pressure deficit the situation which exists between two solutions (↑) on either side of a semipermeable membrane (↑) in which a substance has been added to one of the solutions which cannot pass through the membrane and impedes the passage of water from that solution. The greater the concentration of the solution, the higher the diffusion pressure deficit and the lower the osmotic potential (↑) of that solution.

turgor (*n*) the condition in a plant cell in which water has diffused (↑) into the cell vacuole (p. 11) by osmosis (↑), causing the cell to swell, because the cell fluid was at a lower osmotic potential (↑) than that of its surroundings.

turgid (*adj*) of a plant cell in which the turgor (↑), which is resisted by the elasticity of the cell wall (p. 8), has brought the cell close to bursting and no more water can enter the cell. Turgidity provides the plant with support.

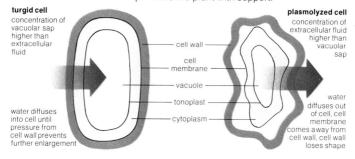

turgid cell
concentration of vacuolar sap higher than extracellular fluid

water diffuses into cell until pressure from cell wall prevents further enlargement

cell wall
cell membrane
vacuole
tonoplast
cytoplasm

plasmolyzed cell
concentration of extracellular fluid higher than vacuolar sap

water diffuses out of cell, cell membrane comes away from cell wall, cell wall loses shape

turgor pressure the pressure exerted by the bulging cell wall (p. 8) during osmosis (↑) into the vacuole (p. 11) of a plant cell.

plasmolysis (*n*) a loss of water, and hence turgidity (↑) from a plant cell when it is surrounded by a more concentrated solution (↑). The cytoplasm (p. 10) loses volume and contracts away from the cell wall (p. 8) causing wilting.

flaccid (*adj*) of cell tissue (p. 83), weak or soft.

wilt (*v*) *of leaves and green stems* to droop.

stoma (*n*) one of the many small holes or pores
(↓) in the leaves (mainly) and stems of plants
through which gas and water vapour exchange
take place. They are able to open and close by
means of their neighbouring guard cells (↓).
stomata (*pl.*).

pore (*n*) a small opening in a surface.

guard cell one of the pair of special, crescent-
shaped cells which surround each stoma (↑)
and which enable the stoma to open or close in
response to light intensity by osmosis (p. 118).
When the guard cells are turgid (p. 119) the
stoma is open.

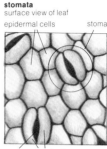

stomata
surface view of leaf
epidermal cells   stoma

pore   guard cells

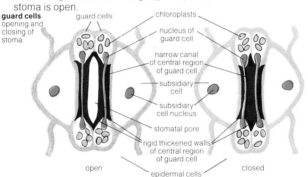

guard cells
opening and
closing of
stoma

guard cells   chloroplasts

nucleus of
guard cell

narrow canal
of central region
of guard cell

subsidiary
cell

subsidiary
cell nucleus

stomatal pore

rigid thickened walls
of central region
of guard cell

open

epidermal cells

closed

substomatal chamber the space below the
stoma (↑).

transpiration (*n*) the loss of water from a plant
through the stomata (↑). It is controlled by the
action of the stomata. It provides a flow
(transpiration stream (p. 122)) of water through
the plant and also has a cooling effect as the
water evaporates from the plant's surface. It is
affected by temperature, relative humidity (↓),
wind speed. As air and leaf temperatures
increase so does the rate of transpiration. The
lower the humidity of the atmosphere, the
faster is the rate of transpiration. Increasing
wind speed normally increases the transpiration
rate provided the cooling effect is not greater.

transpiration
water
evaporates
from
leaves

water
transported
upward in
the vascular
system

water enters
plant through
roots

guttation

high humidity

droplets of
water exuded
from hydathodes
(ends of veins at leaf margin)

**movement of water from
soil to centre of root**

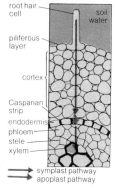

root hair
cell

soil
water

piliferous
layer

cortex

Casparian
strip
endodermis
phloem
stele
xylem

→ symplast pathway
→ apoplast pathway

**symplast and apoplast
pathways**   plasmodesmata

apoplast   **symplast**
substances   substances
translocated   translocated
through cell   through living
walls and inter-   cells and
cellular spaces   plasmodesmata

relative humidity the percentage of water vapour
contained in the air. When the relative humidity
is 100 per cent the air is saturated.

guttation (*n*) the process which takes place in
some plants in conditions of high relative
humidity (↑) in which water is actively secreted
(p. 106) in liquid form by special structures
called hydathodes (found at the end of the
veins on the leaves) rather than being lost as
water vapour. This takes place because of
osmotic (p. 118) absorption (p. 81) of water by
the roots.

atmospheric pressure the pressure which is
exerted on the surface of the Earth by the
weight of the air in the atmosphere.

vacuolar pathway a pathway for the passage of
water by osmosis (p. 118). Vacuoles (p. 11)
contain a fluid with a lower osmotic potential
(p. 118) than water so that the vacuole will take
in water until it becomes turgid (p. 119).

symplast pathway a pathway for the transport of
water through a plant by diffusion (p. 119) from
one cell to the next through the cytoplasm
(p. 10) along the threads called plasmodesmata
(p. 15) which link adjacent cells.

apoplast pathway a pathway for the transport of
water in a plant, particularly across the root
cortex (p. 86), by diffusion (p. 119) along
adjacent cell walls (p. 8).

mass flow a hypothesis (p. 235) developed by
Munch in 1930 to explain the transport of
substances in the phloem (p. 84). Mass flow
takes place in the sieve tube (p. 84) lumina as
water is taken up by osmosis (p. 118) in actively
photosynthesizing (p. 93) regions where
concentration is high and flows to areas where
water is lost as the products of photosynthesis
are being used up or stored and, therefore,
concentration is low. Water is carried in the
opposite direction in the xylem (p. 84) by the
transpiration stream (p. 122).

root pressure the pressure in a plant which
causes water to be transported from the root
into the xylem (p. 84) by the plant's osmotic
(p. 118) gradient.

Casparian strip a thickened waterproof layer which covers the radial and the transverse cell walls (p. 8) of the endodermis (p. 86) so that all water which is transported from the root cortex (p. 86) to the xylem (p. 84) must pass through the cytoplasm (p. 10) of the endodermis cells.

transpiration stream the continuous flow of water which takes place in a plant through the xylem (p. 84) as water is lost to the atmosphere by transpiration (p. 120) and taken up from the soil by the root hairs (p. 81).

cohesion theory the theory (p. 235) which explains that a column of water may be held together by molecular forces of attraction permitting the ascent of sap up a tall stem without falling back or breaking. There is stress or tension in the column of water as water is lost from the xylem (p. 84) vessels by osmosis (p. 118). Similarly, molecular forces of adhesion will cause water to cling to other substances and thereby rise up a stem by capillarity.

translocation[1] (*n*) the transport of organic material through the phloem (p. 84) of a plant. The material includes carbohydrates (p. 17) such as glucose (p. 17), amino acids (p. 21) and plant growth substances (p. 138).

active transport the method by which, with the use of energy, molecules are transported across a cell membrane (p. 14) against a concentration gradient. It probably involves the use of molecular carriers.

transcellular strand hypothesis a hypothesis (p. 235) developed by Thaine to explain why transport rates in plants appear to be greater than would be possible by diffusion (p. 119). It suggests that active transport (↑) takes place along fibrils (p. 11) of protein (p. 21) that pass through the sieve tube (p. 84).

electro-osmotic hypothesis a hypothesis (p. 235) developed by Spanner to explain why transport rates in plants appear to be greater than would be possible by diffusion (p. 119). It suggests that electro-osmotic forces exist across the sieve plates (p. 85).

representations of two theories of phloem transport

flow of sap ▶

transcellular strand hypothesis

fibrils

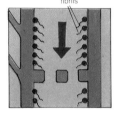

electro-osmotic hypothesis

electro-magnetic force

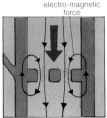

double circulatory system of a mammal

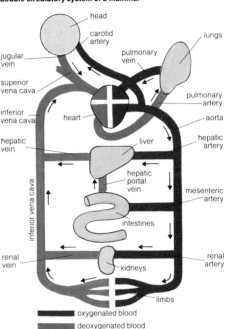

oxygenated blood

deoxygenated blood

single circulation of a fish

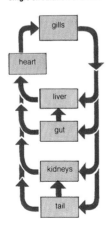

circulatory system a system in which materials
can be transported around the body of an
animal, needed because the volume of the
animal is usually too great for transport to be
effective by diffusion (p. 119).

single circulation a circulatory system (↑), such
as that which occurs in fish, in which the blood
(p. 90) passes through the heart (p. 124) once
on each complete circuit.

double circulation a circulatory system (↑), such
that which occurs in birds and mammals (p. 80), in
which the blood (p. 90) passes through the heart
(p. 124) twice on each complete circuit and so
maintains the system's blood pressure. In this
system, the heart is divided into left and right sides.

open circulatory system a circulatory system
(p. 123), e.g. in arthropods (p. 67), in which the
blood (p. 90) is free in the body spaces for most
of its circulation. The organs lie in the
haemocoel (p. 68) and blood from the arteries
(p. 127) bathes the major tissues (p. 83) before
diffusing (p. 119) back to the open ends of the
veins (p. 127). There are no capillaries (p. 127).

closed circulatory system a circulatory system
(p. 123), e.g. in vertebrates (p. 74), in which the
blood (p. 90) is contained within vessels (p. 127)
for the greater part of its circulation.

heart (n) a muscular (p. 143) organ or specialized
blood (p. 90) vessel (p. 127) which pumps
around the circulatory system (p. 123).

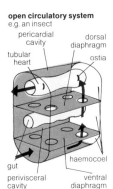

open circulatory system
e.g. an insect

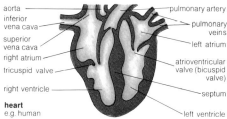

heart
e.g. human

atrium (n) the region or chamber of the heart (↑)
which receives the blood (p. 90). The heart of a
mammal (p. 80) has a left and a right atrium
which are the receivers of oxygenated (p. 126)
blood from the lungs (p. 115) and deoxygenated
(p. 126) blood from the body respectively. Also
known as **auricle**. **atria** (pl.).

ventricle (n) a muscular (p. 143) region or
chamber of the heart (↑) which by regular
contractions pumps the blood (p. 90). The
heart of a mammal (p. 80) has a left and a right
ventricle which pump oxygenated (p. 126)
blood to the body and deoxygenated (p. 126)
blood to the lungs respectively.

cardiac cycle the cycle in which, by rhythmical
muscular (p. 143) contractions, blood (p. 90)
flows into the atria (↑) of the heart (↑) and is
pumped out of the ventricles (↑).

systole (n) contraction phase of cardiac cycle (↑).

diastole (n) relaxation phase of cardiac cycle (↑).

position of heart in various invertebrates

earthworm

crustacean

spider

valve (*n*) a flap or pocket which only allows a liquid, e.g. blood (p. 90), to flow in one direction.

atrioventricular valve the valve which separates the left ventricle (↑) and atrium (↑) preventing blood (p. 90) from flowing back into the atrium by the closure of two membranous (p. 14) flaps. Also known as **mitral valve**.

bicuspid valve = atrioventricular valve (↑).

tricuspid valve the valve which separates the right ventricle (↑) and atrium (↑).

tendinous cords the tough connective tissue (p. 88) in the heart (↑) which prevents the atrioventricular (↑) and tricuspid valves(↑) from turning inside out during contraction.

pocket valves valves between the ventricles (↑) and the pulmonary artery (p. 128) and aorta (↓) which, when closed, prevent the return of blood to the ventricles.

semi-lunar valves = pocket valves.

aorta (*n*) the major artery (p. 127) carrying oxygenated (p. 126) blood (p. 90) from the heart (↑).

myogenic muscle (*n*) muscle (p. 143), e.g. cardiac muscle (p. 143), which may contract without nervous (p. 149) stimulation although its rate of contraction is controlled by such stimulation.

heartbeat (*n*) the rhythmic contraction of the myogenic muscle (↑) of the heart (↑). Also known as **cardiac rhythm**.

sino-atrial node a group of cells in the right atrium (↑) which is responsible for maintaining the heartbeat (↑) by nervous (p. 149) stimulation relayed by it.

pacemaker (*n*) = sino-atrial node (↑).

atrio-ventricular node a second group of cells in the right atrium (↑) which receives the nervous (p. 149) stimulation from the sino-atrial node (↑).

Purkinje tissue nervous (p. 149) tissue (p. 83) which conducts the nervous stimulation from the sino-atricular node (↑) to the tip of the ventricle (↑) ensuring that the ventricle contracts from its tip upwards to force blood (p. 90) out through the arteries (p. 127).

sympathetic nerve a motor nerve (p. 149) which arises from the spinal nerve and releases adrenaline (p. 152) into the heart muscle (p. 143) to increase the heartbeat (↑).

heartbeat action of the heart valves closed

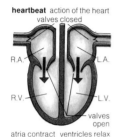

R.A.

L.A.

R.V.

L.V.

valves open

atria contract ventricles relax

valves open

valves closed

ventricles contract atria relax

**changes in volume and pressure
during a mammalian cardiac cycle**

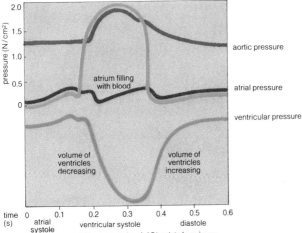

vagus nerve a motor nerve (p. 149) which arises
from the medulla oblongata (p. 156) and
releases acetylcholine (p. 152) into the heart
muscle (p.143) to decrease the heartbeat (p.125).

pulse (*n*) a wave of increased blood (p. 90)
pressure which passes through the arteries (↓)
as the left ventricle (p. 124) pumps its contents
into the aorta (p. 125).

pigment (*n*) a coloured substance. For example,
myoglobin is a variety of haemoglobin (↓) found
in muscle (p. 143) cells, and *chlorocruorin* is a
respiratory pigment containing iron which is
found in the blood (p. 90) of some polychaetes
(p. 65). *See also* chlorophyll (p. 12).

haemoglobin (*n*) a red pigment (↑) and protein
(p. 21) containing iron, which is found in the
cytoplasm (p. 10) of the red blood cells (p. 91)
of vertebrates (p. 74). It combines readily with
oxygen to form oxyhaemoglobin and in this
form oxygen is transported to the tissues
(p. 83) from the lungs (p. 115).

oxyhaemoglobin (*n*) *see* haemoglobin (↑).

oxygenated (*adj*) containing or carrying oxygen.

deoxygenated (*adj*) not oxygenated (↑).

**the pattern of excitation
that accompanies
contraction of the heart**

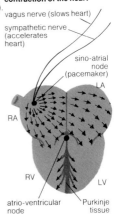

a network of capillaries

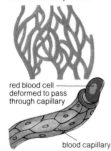

red blood cell
deformed to pass
through capillary

blood capillary

artery

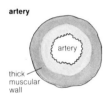

artery

thick
muscular
wall

vein

vein

thin muscular
wall

valve in a vein

valve

haemocyanin (*n*) a blue pigment (↑) and protein
(p. 21) containing copper, which is found in the
plasma (p. 90) of certain invertebrates (p. 75). It
also combines with oxygen to transport it to the
tissues (p. 83).

Bohr effect the effect of increasing the likelihood
of the dissociation of oxygen from oxyhaemoglobin
(↑) as the level of carbon dioxide is increased
so that with increased activity more oxygen is
passed to the body tissues (p. 83).

vascular system the system of vessels (↓) which
transport fluid throughout the body of an organism.

capillary (*n*) any of the very large numbers of tiny
blood vessels (↓) which form a network
throughout the body. They present a large
surface area and are thin-walled to aid gas
exchange (p. 112).

sphincter muscle any of the muscles (p. 143)
which, by contraction, close any of the hollow
tubes, organs, or vessels (↓) in an organism.

vessel[a] (*n*) a channel or duct with walls e.g. blood
(p. 90) flows through a blood vessel.

vein[a] (*n*) any one of the tubular vessels (↑) that
conveys blood (p. 90) back to the heart (p. 124).
Veins are quite large in diameter but thinner
walled than arteries (↓) and the blood is carried
under relatively low pressure. Veins have
pocket valves (p. 125) which ensure that the
blood is carried towards the heart only.

venule (*n*) a small blood vessel (↑) that receives
blood from the capillaries (↑) and then, with
other venules forms the veins (↑).

artery (*n*) any one of the tubular vessels (↑) that
conveys blood (p. 90) from the heart (p. 124).
They are smaller in diameter than veins (↑) but the
walls are thicker and more elastic and the blood is
carried at relatively high pressure. With the exception
of the aorta (p. 125) and the pulmonary artery
(p. 128), arteries have no pocket valves (p. 125).

arteriole (*n*) a small artery (↑).

sinus (*n*) any space or chamber, such as the
sinus venosus which is a chamber found within
the heart (p. 124) of some vertebrates (p. 74)
especially amphibians (p. 77), and lies between
the veins (↑) and the atrium (p. 124).

pulmonary circulation the part of the double
circulation (p. 123) in which deoxygenated
(p. 126) blood (p. 90) is pumped from the heart
(p. 124) to the lungs (p. 115).

pulmonary artery the artery (p. 127) which
carries the deoxygenated (p. 126) blood (p. 90)
pumped from the heart (p. 124) to the lungs (p.115).

pulmonary vein the vein (p. 127) which carries
the oxygenated (p. 126) blood (p. 90) from the
lungs (p. 115) back to the heart (p. 124).

systemic circulation the part of the double
circulation (p. 123) in which blood (p. 90) is
pumped from the heart (p. 124) throughout the
body of the animal.

arterio-venous shunt vessel a small blood vessel
(p. 127) which bypasses the capillaries (p. 127)
and carries blood (p. 90) from the arteries (p.127)
to the veins (p. 127) and therefore regulates the
amount of blood which enters the capillaries.

lymph (*n*) a milky or colourless fluid which drains
from the tissues (p. 83) into the lymphatic
vessels (↓) and is not absorbed (p. 81) back
into the capillaries (p. 127). It is similar to tissue
fluid and contains bacteria (p. 42) but does not
contain large protein (p. 21) molecules.

lymphatic vessel any one of the vein-like (p. 127)
vessels (p. 127) that carry lymph (↑) from the
tissues (p. 83) into the large veins that enter the
heart (p. 124).

lymph node a swelling in the lymphatic vessel
(↑), especially in areas such as the groin or
armpits, which contain special white blood
cells (p. 91) known as macrophages (p. 88).

platelet (*n*) any of the fragments of cells present
in the blood (p. 90) plasma (p. 90) which are
formed in the red bone marrow (p. 90) and
which prevent bleeding by combining at the
point of an injury and releasing a hormone
(p. 130) which stimulates blood clotting (↓).
They also release other substances which
cause blood vessels (p. 127) to constrict so
that they also prevent capillary (p 127) bleeding.

coagulate (*v*) = clot (↓).

anticoagulant (*n*) a substance that stops blood
(p. 90) clotting (↓).

lymphatic system

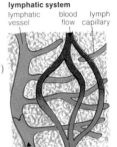

lymphatic vessel — blood flow — lymph capillary

lymph flow — blood flow

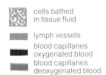

cells bathed in tissue fluid

lymph vessels

blood capillaries oxygenated blood

blood capillaries deoxygenated blood

platelet

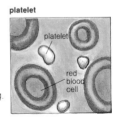

platelet

red blood cell

clot (*v*) *of liquids* to become solid, for example, blood (p. 90) clots in air. **clot** (*n*).

blood groups in humans, there is a system of multiple alleles (p. 205) which gives rise to four main different blood groups with different antigens (p. 234) or proteins (p. 21) on the surface of the red blood cells (p. 91). The A allele, B allele and O allele (producing no antigens) may combine to give any of the following blood group combinations: AA, AO, BB, BO, AB or OO. A and B alleles are both dominant (p. 197) to O so that there are four groups, A, B, O and AB.

rhesus factor an antigen (p. 234) which is present in the blood (p. 90) of rhesus monkeys and most, but not all, humans. During pregnancy (p. 195) or following transfusion of blood containing the rhesus factor (Rh+) into blood lacking it (Rh–), breakdown of the red blood cells (p. 91) can occur with dangerous results.

the four main blood groups

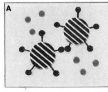
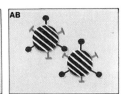

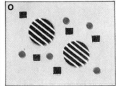

●— **A** antigen —| **B** antigen ■ **A** antibody ● **B** antibody

	blood group	antigens on red cells	antibodies in serum	can receive blood type	can donate blood to
	A	**A**	**B**	groups **A** and **O**	groups **A** and **AB**
	B	**B**	**A**	groups **B** and **O**	groups **B** and **AB**
universal recipients	**AB**	**A** and **B**	none	groups **A, B, AB** and **O**	group **AB**
universal donors	**O**	none	**A** and **B**	groups **O** only	groups **A, B, AB** and **O**

homeostasis (*n*) the maintenance of constant internal conditions within an organism, thus allowing the cells to function more efficiently, despite any changes that might occur in the organism's external environment (p. 218).

endocrine system a system of glands (p. 87) in animals which produce hormones (↓). This system and the nervous system (p. 149) combine to control the functions of the body.

endocrine gland a gland (p. 87) which produces hormones (↓).

hormone (*n*) a substance made in very small amounts in one part of an organism and transported to another part where it produces an effect. (1) In plants, the hormones can be referred to as growth substances (p. 138). (2) In animals, hormones are secreted (p. 106) by the endocrine glands (↑) into the blood stream (p. 90), where they circulate to their destination.

adrenal glands in mammals (p. 80), a pair of endocrine (↑) glands (p. 87) near the kidneys (p. 136). They are divided into two parts; the *medulla*, the inner part which secretes (p. 106) adrenaline (p. 152) and noradrenaline (p. 152), and the *cortex*, the outer part which secretes various steroid (p. 21) hormones (↑).

homoiothermic (*adj*) of an organism which maintains its body temperature at a constant level in changing external circumstances. These organisms, including mammals (p. 80), for example, are usually regarded as 'warm blooded' because their body temperature is usually above that of the surroundings.

endothermic (*adj*) = homoiothermic (↑).

poikilothermic (*adj*) of an organism whose body temperature varies with and is roughly the same as that of the environment (p. 218). These organisms, not including birds and mammals (p. 80), are usually regarded as 'cold blooded' although their body temperature may be higher or lower than that of the environment depending on such factors as wind speed or the sun's radiation (↓). They have a lower metabolic rate (p. 32) as their body temperature falls.

exothermic (*adj*) = poikilothermic (↑).

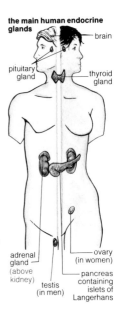

the main human endocrine glands

brain

pituitary gland

thyroid gland

adrenal gland (above kidney)

ovary (in women)

testis (in men)

pancreas containing islets of Langerhans

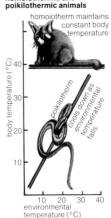

comparison of homoiothermic and poikilothermic animals

homoiotherm maintains constant body temperature

poikilotherm

cools down as environmental temperature falls

body temperature (°C)

10 20 30 40
environmental temperature (°C)

**heat gains and losses in a
reptile (poikilothermic)
animal**

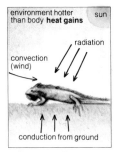

environment cooler than body
heat losses

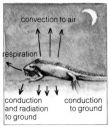

radiation (*n*) the transfer of heat from a hot
object, such as the sun, to a cooler object,
such as the earth or the body of an organism,
through space without increasing the
temperature of the space.

evaporation (*n*) the change from a liquid to a vapour
or gas that takes place when the liquid is
warmed to a temperature at or below its boiling
point.

conduction (*n*) the transfer of heat through a solid.

convection (*n*) the transfer of heat in a fluid as
the warmed portion of the fluid rises and the
cool portion falls.

hair (*n*) a single-celled or many celled outgrowth
from the dermis (↓) of a mammal (p. 80) made
up of dead material and including the substance
keratin. Among other functions, a coat of hair
insulates the mammal's body from excessive
warming or cooling especially if the hairs are
raised to trap a layer of insulating air around
the body. Hair is a characteristic of mammals.

skin (*n*) the outer covering of an organism which
insulates it from excessive warming or cooling,
prevents damage to the internal organs,
prevents the entry of infection, reduces the loss
of water, protects it from the sun's radiation (↑)
and it also contains sense organs which make
the organism aware of its surroundings.

dermis (*n*) = skin (↑).

epidermis (*n*) the outer layer of the skin (↑). The
epidermis is made up of three main layers of
cells: the continuous *Malpighian layer* is able to
produce new cells by division and so replace
epidermal layers as they are worn away; the
granular layer grades into the harder, outer
cornified layer which is composed of dead
cells only and forms the main protective part of
the skin. This cornified layer may become very
hard and thick in areas that are constantly
subject to wear, such as the soles of the feet.

sebaceous gland one of the many glands (p. 87)
contained within the skin (↑) that open into the
follicles (p. 132) and which secrete (p. 106) an
oily antiseptic substance which repels water
and keeps the skin flexible.

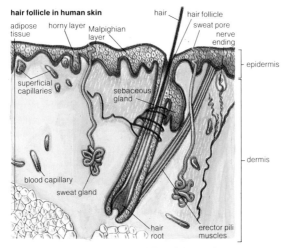

hair follicle in human skin

hair — hair follicle
adipose tissue — horny layer — Malpighian layer — sweat pore — nerve ending
epidermis
superficial capillaries — sebaceous gland
dermis
blood capillary
sweat gland
hair root — erector pili muscles

hair follicle (*n*) a pit within the Malphigian and granular layers of the epidermis (p. 131) which contains the root of a hair and from which the hair grows by cell division. Sebaceous glands (p. 131) open into the hair follicles and also contain muscles (p. 143) to erect the hairs for additional insulation as well as nerve (p. 149) endings for sensitivity.

sweat gland a coiled tubular gland (p. 87) within the epidermis (p. 131) which absorbs moisture containing some salts and minerals from surrounding cells and releases it to the surface through a tube causing the skin (p. 13) to be cooled as the moisture evaporates into the atmosphere.

hibernation (*n*) a process by which certain organisms respond to very low external temperatures in a controlled fashion. The core temperature of the animal falls to near that of the environment (p. 218) and there is a resulting drop in the metabolic rate (p. 32) although the nervous system (p 149) continues to operate so that should the temperature fall near to the lethal temperature, the animal increases its metabolic rate to cope with it.

hyperthermia (*n*) overheating. The condition in which, as a result of vigorous activity, disease, or heating by radiation (p. 131), the body temperature of an organism rises above its normal level. As a result of nerve impulses (p. 150) sent to the hypothalmus (p. 156) counter measures are taken including dilation of the blood vessels (p. 127) to allow greater heat loss by radiation, convection (p. 131) and conduction (p. 131), and sweating to cool the body surface as moisture evaporates.

hypothermia (*n*) overcooling. The condition in which the temperature of an organism falls below its normal level. As a result of nerve impulses (p. 150) sent to the hypothalamus (p. 156) counter measures are taken including a reduction in sweating, constriction of the blood vessels (p. 127) to reduce the amount of heat being lost by radiation (p. 131), conduction (p. 131) and convection (p. 131), rapid spasmodic contraction of the muscles (p. 143) to cause shivering, and an increase in metabolic rate (p. 32).

aestivation (*n*) (1) the condition of inactivity or torpor into which some animals enter during periods of drought or high temperatures. Lung fish, for example, bury themselves in mud at the start of the dry season and re-emerge when the rain begins to fall again. (2) the arrangement of parts in a flower bud.

leaf fall the condition into which some plants enter during periods of extreme water shortage by losing some of their leaves to reduce water losses by transpiration (p. 120).

osmoregulation (*n*) the process by which an organism maintains the osmotic potential (p.118) in its body fluids at a constant level e.g. freshwater fish take in large volumes of water through the gills (p.113) by osmosis (p.118) which are then excreted (p.134) as urine (p.135) from the kidneys (p.135). Marine fish, either drink sea water (bony fish (p.76)) so that salts are absorbed (p. 81) by the gut (p.98) and water then follows by osmosis, with the salts eliminated by the gills, or retain urea (p.134) so that their fluids are hypertonic (p. 118) to sea water and then, like freshwater fish, they take in more water through their gills.

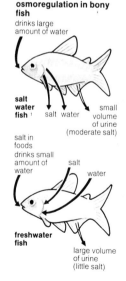

osmoregulation in bony fish

drinks large amount of water

salt water fish

salt water

small volume of urine (moderate salt)

salt in foods
drinks small amount of water

salt

water

freshwater fish

large volume of urine (little salt)

carotid body a small oval structure in the carotid artery (p. 127) containing nerves (p. 149) which respond to the oxygen and carbon dioxide content of the blood (p. 90) and so control the level of respiration (p. 112).

carotid sinus a small swelling in the carotid artery (p. 127) containing nerves (p. 149) which respond to blood (p. 90) pressure and so control circulation (p. 123).

excretion (*n*) the process by which the waste and harmful products of metabolism (p. 26) such as water, carbon dioxide, salts and nitrogenous compounds, are eliminated from the organism.

chemical formulae of nitrogenous wastes

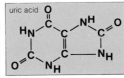

excretory organs in invertebrates
Malpighian tubules of insec

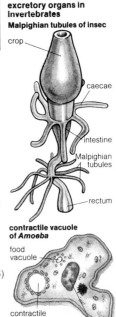

urea (*n*) a nitrogenous organic compound (p. 15) which is soluble in water and which is the main product of excretion (↑) from the breakdown of amino acids (p. 21) in certain animals.

uric acid a nitrogenous organic compound (p. 15), insoluble in water, which is the main product of excretion (↑) from the breakdown of amino acids (p. 21) in certain animals. Because it is insoluble, uric acid is not toxic and can be excreted without losing large amounts of water.

ammonia (*n*) a nitrogenous inorganic compound (p. 15) which is very toxic and may only be found as a product of excretion (↑) in organisms where large amounts of water are available for its removal, such as in aquatic animals.

contractile vacuole a vacuole (p. 11) present in the endoplasm (p. 44) of the Protozoa (p. 44) which is important in the osmoregulation (p. 133) of these organisms. In hypotonic (p.118) solutions (p. 118) the vacuole swells and then seems to contract, releasing to the exterior, water which has entered the cell with food or from the surroundings. The water passes into the solution. In hypertonic (p. 118) and isotonic (p. 118) solutions the vacuole disappears.

contractile vacuole of Amoeba

Malpighian tubule one of a number of narrow, blind tubes which are the main organs of osmoregulation (p. 133) and excretion (↑) in insects (p. 69) and other members of the Arthropoda (p. 67). They arise from the gut (p. 98) and in them, uric acid (↑) crystals are produced which can be eliminated with little water loss. Although insects do possess a thin, waterproof cuticle (p. 145) water is still lost through the joints (p. 146) and by respiration (p. 112).

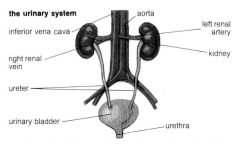

the urinary system

aorta

inferior vena cava

left renal artery

right renal vein

kidney

ureter

urinary bladder

urethra

ureter (*n*) the tube or duct which carries urine (↓) from the kidney (p. 136) to the bladder (↓).

bladder[a] (*n*) an extensible sac into which the ureter (↑) passes which fills with urine (↓) secreted (p. 106) continuously by the kidneys (p. 136). When full, it is opened by a sphincter muscle (p. 127), contracts, and releases the urine.

urine (*n*) the fluid which is finally expelled from the kidneys (p. 136). It contains urea (↑), or uric acid (↑) together with other materials and water.

anti-diuretic hormone a hormone (p. 130) synthesized by the hypothalamus (p. 156) and secreted (p. 106) by the pituitary gland (p. 157). It increases the reabsorption of water in the kidney (p. 136) tubules thus increasing the concentration of the urine (↑).

aldosterone (*n*) a hormone (p. 130) secreted (p. 106) by the adrenal glands (p. 130) which stimulates the reabsorption of sodium from the kidneys (p. 136) and increases the excretion (↑) of potassium thus tending to increase the concentration of sodium in the blood (p. 90) while decreasing the concentration of potassium.

kidney (*n*) in mammals (p. 80), one of a pair of organs which form the main site of excretion and may also be involved in osmoregulation (p. 113). Blood (p. 90) is pumped by the heart (p. 124) under pressure through the kidneys and, by reabsorptions and secretions (p. 106), useful substances are returned to the blood while wastes are eliminated in the urine. (↓).

nephron (*n*) one of the main structures of excretion (p. 134) in the kidneys (↑). It is a microscopic (p. 9) tubule which is made up of a Malpighian corpuscle and a drainage duct. It is in the nephrons that the processes of filtration and reabsorption take place.

kidney

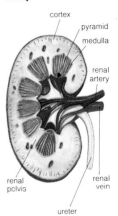

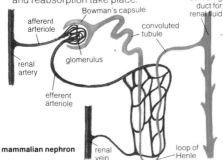

mammalian nephron

Bowman's capsule part of the Malpighian corpuscle in the nephron (↑). It is a cup-shaped swelling surrounding the glomerulus (↓) forming part of the structure through which blood (p. 90) is forced under pressure and by which it is 'purified' by a process known as ultra-filtration (↓).

loop of Henle a U-shaped tubule in the nephron (↑) into which isotonic (p. 118) renal fluid (↓) is pumped. As the fluid passes in the counter direction along the other arm of the tubule, sodium ions are actively transported (p. 122) across into the first arm so that there is a high concentration of sodium ions in the bend of the tube. From here there are collecting ducts which open into the renal pelvis from where water is drawn out by osmosis (p. 118). Thus, in the collecting ducts, the renal fluid becomes hypertonic (p. 118).

glomerulus (*n*) a knot of capillaries (p. 127) which form part of the Malpighian corpuscle and into which blood (p. 90) is pumped via the renal artery (p. 127) and arterioles (p. 127). Blood is forced through the walls of the capillaries into the Bowman's capsule (↑).

renal fluid a fluid consisting mostly of blood plasma (p. 90) and the soluble materials contained within it, which, the process of ultrafiltration (↓) passes through the kidneys (↑).

ultrafiltration (*n*) the process by which a large proportion of the blood plasma (p. 90) and the soluble materials contained within it are forced under pressure through the walls of the glomerulus (↑), through the walls of the Bowman's capsule (↑) and into the lumen (↓) of the nephron (↑).

lumen (*n*) a space within a tube or sac.

mesophyte e.g. beech

sheds leaves in autumn

hydrophyte e.g. *Nuphar*

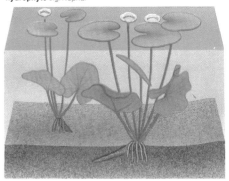

xerophyte
e.g. cactus

thick succulent stems with thick cuticle

hot dry air

spines

dry desert sand

hydrophyte (*n*) a plant which is adapted to grow in water or in very wet conditions. Leaves and stems often contain air spaces to aid the flotation of the whole plant or part of it.

mesophyte (*n*) a plant which is adapted to grow in habitats (p. 217) with normal supplies of water. Typically, they have large, flattened leaves that are lost during leaf fall (p. 133).

xerophyte (*n*) a plant which is adapted to grow in very dry environments (p. 218).

growth substance (*n*) a hormone (p. 130) which, in very small quantities, can increase, decrease, or otherwise change the growth of a plant or part of a plant.

indole-acetic acid IAA. A growth substance (↑) which causes plant cells to grow longer and causes cells to divide. IAA is the most common of the group of growth substances called auxins.

auxin (*n*) *see* indole-acetic acid (↑).

indole-acetic acid (IAA)

gibberellin (*n*) a growth substance (↑) which may effect the elongation of cells in stems and which may cause the area of leaves to increase. They also promote a variety of effects in plant growth such as seed germination (p. 168), flowering and the setting of fruit.

gibberellin
e.g. gibberellic acid 1

cytokinin (*n*) a growth substance (↑) which, in association with IAA, affects the rate of cell division, promotes the formation of buds, and is essential for the growth of healthy leaves. Also known as **kinin**.

cytokinins e.g. kinetin

absicin (*n*) a growth substance (↑) which inhibits plant growth, prevents germination (p. 168), and tends to promote buds to become dormant. Absicin seems to work against normal growth substances by preventing the manufacture of proteins (p. 21) etc.

ethene (*n*) a growth substance (↑) which is produced as the result of normal metabolism (p. 26) in plants and which may cause leaves to fall and fruit to ripen. Also known as **ethylene**.

florigen (*n*) a growth substance (↑) which, although it has never been isolated, is believed to promote the production of flowers.

tropism (*n*) the way in which the direction of growth of a plant responds to external stimuli. **trophic** (*adj*).

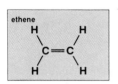
ethene

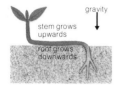
geotropism
gravity
stem grows upwards
root grows downwards

geotropism (*n*) a tropism (↑) in which the various parts of a plant grow in response to the pull of the Earth's gravity. For example, primary roots grow downwards and are referred to as being positively geotrophic while the main stems grow upwards and are referred to as being negatively geotrophic.

statolith (*n*) a large grain of starch (p. 18) which is found in plant cells and which is thought to respond to the effects of gravity causing the effects of geotropism (p. 139).

phototropism (*n*) a tropism (p. 139) in which various parts of the plant grow in response to the direction from which light is falling on the plant. Stems tend to grow towards the light and are referred to as being positively phototrophic. The roots of some plants e.g. those of climbers grow away from the source of light and are referred to as being negatively phototrophic.

phototropism

auxin in phototropism

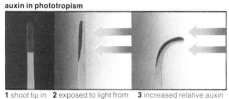

1 shoot tip in dark, auxin evenly concentrated   **2** exposed to light from one side, auxin concentration increases on dark side and decreases on light side   **3** increased relative auxin concentration on dark side causes cells on dark side to elongate, and the shoot bends towards the light

hydrotropism (*n*) a tropism (p. 139) in which the roots of plants grow towards a source of water. Hydrotropism will usually override the effects of geotropism (p. 139). **hydrotrophic** (*adj*).

chemotropism (*n*) a tropism (p. 139) in which the roots of a plant or the hyphae (p. 46) of a fungus (p. 46) may grow towards a source of food materials. **chemotrophic** (*adj*).

thigmotropism (*n*) a tropism (p. 139) in which, by the stimulus of touch, certain parts of particular plants, such as the stems of climbing plants, may coil around a support. **thigmotrophic** (*adj*).

nastic movements growth movements, such as the opening and closing of flowers, which although they occur as a result of external stimuli, such as the presence or absence of light, do not take place in a particular direction.

photonasty (*n*) a nastic movement (↑) which is a response to the presence or absence of light or even to light levels e.g. the flowers of daisies close at night and only open during the daylight.

thigmonastic movements of the leaves of a 'sensitive plant' after it has been touched

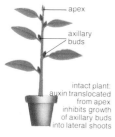

apical dominance

apex

axillary buds

intact plant: auxin translocated from apex inhibits growth of axillary buds into lateral shoots

apex removed: lateral shoots grow

thermonasty (*n*) a nastic movement (↑) which is a response to the surrounding temperature. For example, the flowers of some plants will open when the weather is warm.

thigmonasty (*n*) a nastic movement (↑) in which the response is to touch. For example, the leaves of the South American plant, commonly known as the 'sensitive plant', fold back when touched.

taxic movements the movement of an organism in which the response takes place in relation to the direction of the stimulus.

phototaxis (*n*) a taxic movement (↑) in which the movement may be away from or towards the direction from which the light is coming. For example, certain insects (p. 69) may hide from the light and are referred to as negatively phototaxic while many algae (p. 44) will move towards the light and are described as positively phototaxic.

thermotaxis (*n*) a taxic movement (↑) in which the movement may be away from or towards regions of higher or lower temperature. For example, a mammal (p. 80) may seek the shade of a tree during the heat of the day to prevent overheating.

chemotaxis (*n*) a taxic movement (↑) in which an organism may move towards a chemical stimulus. For example, a spermatozoid (p. 178) may swim towards a female organ which secretes (p. 106) a substance, such as sucrose (p. 18).

hygroscopic movements movements which take place as the parts of organisms dry out and thicker parts move differently from thinner parts.

autonomic movements movements which take place in an organism without any external stimulus. The stimulus comes from within the organism itself and may include movements such as the coiling of the tendrils of climbing plants such as peas.

apical dominance the state which may occur in plants in which the bud at the tip of a plant stem grows but the lateral ones do not. If the apical bud is removed, lateral branches may grow.

vernalization (*n*) the process whereby certain plants, such as cereal crops, need to be subjected to low temperatures, such as that which occurs through overwintering, during an early part of their growth before they will be induced to flower.

phytochrome (*n*) a light-sensitive pigment (p. 126) present in plant leaves which exists in two forms that can be converted from one to the other. One absorbs red light, the other far red light. In the absence of light, the latter slowly changes back to the former. Phytochromes initiate the formation of hormones (p. 130).

etiolation (*n*) plant growth which takes place in the absence of light. The plants may lack chlorophyll (p. 12) so that they will be yellow or even white in colour. The leaves will be reduced in size and the stems tend to grow much longer.

photoperiodism (*n*) the process in which certain activities, such as flowering or leaf fall (p. 133), respond to seasonal changes in day length.

long-day plants plants, such as cucumber, which only usually flower during the summer months in temperate climates when the hours of daylight exceed about fourteen in twenty-four.

short-day plants plants, such as chrysanthemums, which only usually flower during the spring or autumn months in temperate climates when the hours of daylight are less than about fourteen in twenty-four.

day-neutral plants plants, such as the pea, in which the hours of daylight have no influence on the flowering period.

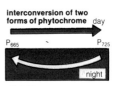

interconversion of two forms of phytochrome

day

P_{665} P_{725}

night

etiolation

etiolated
young plant grown in dark

hooked apex

no chlorophyll

long shoot

little leaf development

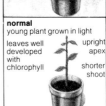

normal
young plant grown in light

leaves well developed with chlorophyll

upright apex

shorter shoot

photoperiodism

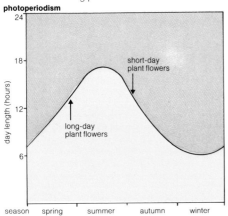

day length (hours)

24

18

12

6

short-day plant flowers

long-day plant flowers

season   spring   summer   autumn   winter

long-day plant

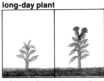

kept under short days

kept under long days

short-day plant

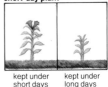

kept under short days

kept under long days

fibre of voluntary muscle

striped
band

structure of striated muscle

endomysium

nucleus

A band
I band

H-line
Z-line
myofibril

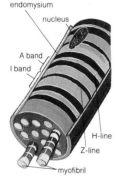

structure of cardiac muscle

connective   branched
tissue   fibre

sarcolemma

intercalated   nucleus
disc

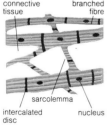

locomotion (*n*) the ability of an organism to move all or part of its body independent of any outside force. An animal can usually move its whole body whereas a plant may only be able to move certain parts, such as petals (p. 179) or leaves, in response to changes in the environment (p. 218).

muscle (*n*) tissue (p. 83) which is made up of cells or fibres that are readily contracted.

fibre (*n*) a thread-like structure.

skeletal muscle muscle (↑) tissue (p. 83) consisting of elongated cells with many nuclei (p. 13) and cross striations in the cytoplasm (p. 10). It usually occurs in bundles and is under voluntary control of the central nervous system (p. 149) so that it contracts when stimulated to do so. These muscles are attached to parts of the skeleton (p. 145) and their contractions cause these parts to move. Skeletal muscle which has a striped look is known as striated muscle. This consists of long, narrow muscle fibres bounded by a membrane (p. 14) and containing many nuclei. The muscle fibres are bound together into bundles. They contract when stimulated.

voluntary muscle = skeletal muscle (↑).

striated muscle = skeletal muscle (↑). Also known as **striped muscle**.

unstriated muscle = involuntary muscle (↓).

involuntary muscle the muscle (↑) which is found in the internal organs and blood vessels (p. 127) and consists of simple tubes or sheets. It is under the involuntary control of the autonomic nervous system (p. 155). Also known as **smooth muscle**.

visceral muscle a smooth or unstriated (↑) muscle (↑) tissue (p. 83) made up of elongated cells held together by connective tissue (p. 88) and activated involuntarily. It is found in all internal organs as well as blood vessels (p. 127) with the exception of the heart (p. 124).

cardiac muscle muscle (↑) tissue (p. 83) found only in the heart (p. 124) walls. It consists of fibres containing cross striated (↑) myofibrils (p. 144). It contracts rhythmically and automatically (i.e. without nervous (p. 149) stimulation).

muscle fibre the elongated cells which make up striated muscles (p. 143) and which consist of a number of myofibrils (↓).

myofibril (*n*) the very fine threads which make up the muscle fibres (↑) and are found in smooth, striated (p. 143), and cardiac muscles (p. 143). They contain the contractile proteins (p. 21) myosin (↓), actin (↓) and tropomyosin (↓).

sarcomere (*n*) the part of the myofibril (↑) which is responsible for the contraction. It is made up of a dark central A band composed of myosin (↓) on either side of which are I bands composed of actin (↓). Each sarcomere is joined to the next by the Z membrane (p. 14). During contraction the I band shortens while the A bands stay more or less the same length so that the muscle filaments slide between one another.

thick filaments the filaments of a myofibril (↑) which are composed of myosin (↓).

thin filaments the filaments of a myofibril (↑) which are composed of actin (↓).

actin (*n*) the contractile protein (p. 21) which comprises one of the main elements in muscle (p. 143) myofibrils (↑). When stimulated actin and myosin (↓) join together to form actomyosin (↓).

myosin (*n*) the contractile protein (p. 21) which comprises the most abundant element in muscle (p. 143) myofibrils (↑). When stimulated actin (↑) and myosin join together to form actomyosin (↓).

actomyosin (*n*) a complex of the two proteins (p. 21) actin (↑) and myosin (↑) which, when they interact to form the complex, result in the contraction of the muscle (p. 143).

tropomyosin (*n*) the third protein found in myofibrils (↑) which may be responsible for controlling the contractions of muscle (p. 143).

sliding filament hypothesis the theory (p. 235) which suggests that when a muscle (p. 143) contracts, the individual filaments do not shorten but that they slide between one another because it can be seen under the electron microscope (p. 9) that the pattern of striations (p. 143) changes during the contraction.

sliding filament hypothesis

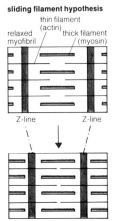

relaxed myofibril

thin filament (actin) thick filament (myosin)

Z-line   Z-line

contracted myofibril

sarcoplasmic reticulum a smooth endoplasmic reticulum (p. 11) which is responsible for absorbing (p. 81) the calcium that is necessary for muscle (p. 143) contraction.

muscle spindle a modified muscle fibre (↑) which is receptive to stimulation and controls the way in which a muscle (p. 143) contracts.

skeleton (*n*) a supporting structure. It can be jointed (p. 146) and the joints (p. 146) are connected by muscles (p. 143) that, when they contract against the limbs as levers, enable the animal to operate the limbs, e.g. the legs, allowing movement on land.

exoskeleton (*n*) the external skeleton (↑) of organisms, such as insects (p. 69), which provides protection for the internal organs and is the structure to which the muscles (p. 143) are attached. For example, the shell of a cockle would be referred to as an exoskeleton.

apodeme (*n*) one of a number of projections on the inside of the exoskeleton (↑) where there are joints (p. 146) and to which the muscles (p. 143) for the movement of those joints are attached.

cuticle[a] (*n*) the outer layer of the endoskeleton (↓) which in animals, such as insects (p. 69), acts as the skeleton (↑) itself. It may be composed of chitin (p. 49) but in shellfish may be hardened with lime-rich salts. It is secreted (p. 106) by the epidermis (p. 131) and is non-cellular. The outer epicuticle is waxy and waterproof in insects and other arthropods (p. 67).

hydrostatic skeleton a form of skeleton (↑) found in soft-bodied animals such as earthworms (p. 66) in which the body fluids themselves provide the structure against which the muscles (p. 143) act.

endoskeleton (*n*) a bony or cartilaginous (p. 90) structure contained within the body of vertebrates (p. 74) which is usually jointed (p. 146) to allow movement and to which the muscles (p. 143) are attached to provide the mechanisms for movement.

musculo-skeletal system the system which enables the animal to move by providing a jointed (p. 146) skeleton (↑) against which the muscles (p. 143) can act to cause operation of the joints using the limbs as levers.

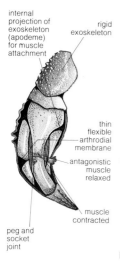

claw of crab showing exoskeleton

internal projection of exoskeleton (apodeme) for muscle attachment

rigid exoskeleton

thin flexible arthrodial membrane

antagonistic muscle relaxed

muscle contracted

peg and socket joint

swimming in fish waves of lateral undulations pass from the head back along the body

swimming (n) the process by which an organism such as a fish propels itself through or on the surface of the water by the action of fins (p. 75) or flexing of the whole body. In fish, when the muscles (p. 143) contract, the body cannot shorten so that it moves from side to side to provide propulsion.

caudal fin the tail fin. The main organ by which a fish propels itself through the water. It is a membrane (p. 14) supported by fin rays attached to the vertebral column (p. 74) of the fish

myotome muscle one of a number of blocks of striated muscle (p. 143) that completely enclose the vertebral column (p. 74) of fish.

joint (n) the region at which any two or more bones of a skeleton (p. 145) come into contact. For example, the elbow joint in a human.

ball and socket joint a movable joint (↑) between limbs in which one bone terminates in a knob-shaped structure which fits into a cup-shape in the meeting bone allowing some movement in all directions, e.g. the joint between the femur and socket of the pelvic girdle (↓).

hinge joint a movable joint (↑) between bones in which the movement can take place in one plane or direction only e.g. the knee joint.

pivot hinge a movable joint (↑) between bones in which movement can take place in all directions by a twisting or rotating movement. For example, the joints in the neck.

ligament (n) the strong, elastic connective tissue (p. 88) which, for example, holds together the limb bones at a joint (↑) and which helps to control the movement of the joint.

tendon (n) the thick cord of connective tissue (p. 88) which connects a muscle (p. 143) to a bone. It is non-elastic so that when the muscle contracts, it pulls against the bone forcing it to move at the joint (↑).

cross sections of fish body

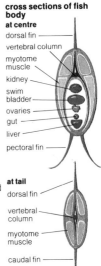

at centre
dorsal fin
vertebral column
myotome muscle
kidney
swim bladder
ovaries
gut
liver
pectoral fin

at tail
dorsal fin
vertebral column
myotome muscle
caudal fin

structure of a joint

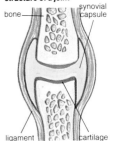

bone
synovial capsule
ligament
cartilage

joints

ball and socket joint socket

hinge joint

ball

muscles of hind leg of a mammal

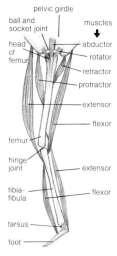

pelvic girdle
ball and socket joint
muscles
head of femur
abductor
rotator
retractor
protractor
extensor
flexor
femur
hinge joint
extensor
tibia-fibula
flexor
tarsus
foot

vertebrae
vertebral column (backbone)

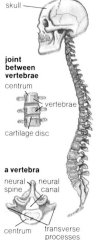

skull

joint between vertebrae
centrum
vertebrae
cartilage disc

a vertebra
neural spine
neural canal
centrum
transverse processes

protractor muscle a muscle (p. 143) which on contraction draws the limb bone forwards.

retractor muscle a muscle (p. 143) which on contraction draws the limb bone backwards.

adductor muscle a muscle (p. 143) which on contraction draws the limb bone inwards.

abductor muscle a muscle (p. 143) which on contraction draws the limb bone outwards.

rotator muscle a muscle (p. 143) which on contraction rotates a limb bone outwards or inwards.

flexor muscle a muscle (p. 143) which on contraction draws two limb bones together.

extensor muscle a muscle (p. 143) which on contraction draws two limb bones apart.

vertebra (n) one of a number of bones, or in some cases, segments of cartilage (p. 90), which make up the vertebral column (p. 74). Each vertebra is usually hollow and has muscles (p. 143) attached to it.

pelvic girdle the part of the skeleton (p. 145) of a vertebrate (p. 74) to which the hind limbs are attached. It is rigid and provides the main support for the weight of the body

limb (n) any part of the body of an animal, apart from the head and the trunk, including, for example, the arms, the legs or the wings.

flight (n) that form of locomotion (p. 143) such as is found in most birds and many insects, whereby the animal is borne through the air either by gliding on the wind using outstretched membranes (p. 14) or by using the lift generated by the special shape and the power provided by wings.

feather (n) one of a very large number of structures which provide the body covering of birds. They distinguish birds from all other animals. Feathers insulate the bird s body, repel water, streamline it, and aid in the power of flight.

down (n) the soft, fluffy feathers (↑) which form the initial covering of very young birds and which are also found on the undersides of adult birds. The individual barbs (p. 148) are not joined together so that down provides better insulation than flight feathers (p. 148) by trapping a layer of air close to the bird's body.

flight feather one of a number of feathers (p. 147) giving birds their streamlined effect and which are elongated to provide the flight surface.

shaft (*n*) the central rod-like but flexible stem of a feather (p. 147) which is the continuation of the quill (↓) that is attached to a feather (like hair) follicle (p. 132) in the skin of a bird. The skin of a bird has no sweat glands (p. 132).

quill (*n*) the hard tube-like part of the feather (p. 147) that is attached to the feather follicle (p. 132) and which is also connected by muscles (p. 143) that are able to alter the angle at which the feather lies in relation to the body of the bird. For example, in cold weather, the bird would raise its feathers to trap extra layers of air for more efficient insulation.

vane (*n*) the flat, blade-like part of the feather (p. 147) which is composed of the shaft (↑) and its attached web of barbs (↓) and barbules (↓).

barb (*n*) a hook-like process which projects from the shaft (↑) of a feather (p. 147). Barbs are interlocked by the barbules (↓).

barbule (*n*) one of the tiny barbs (↑) attached to the barb of a feather (p. 147) and which link the barbs by a system of hooks and troughs to make up the web or vane (↑) of the feather.

pectoralis muscle one of the large, powerful muscles (p. 143) which pull on the wings of a bird to force it upwards and downwards providing the power for flight. Pectoralis muscles are attached to the sternum (↓) of the bird The pectoralis major is the muscle which depresses the wing while the pectoralis minor is responsible for raising it.

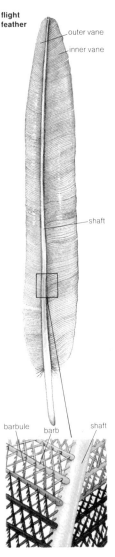

flight feather

outer vane

inner vane

shaft

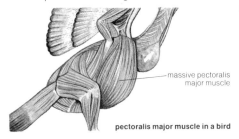

massive pectoralis major muscle

pectoralis major muscle in a bird

barbule   barb   shaft

sternum (*n*) the breast bone of tetrapods.

gliding (*n*) the process of flight (p. 147) whereby the animal holds the wings outstretched so that they function as aerofoils and the animal soars on a cushion of supporting air.

irritability (*n*) the ability of an organism to respond to changes in its environment (p. 218) e.g. the movement of animals in response to noise or being touched.

nervous system the system within the body of an organism which permits the transmission of information through the body so that its various parts are able rapidly to respond to any stimuli.

central nervous system CNS. That part of the nervous system (↑), which in vertebrates (p. 74) includes the brain (p. 155) and the spinal cord (p. 154) which receives nerve impulses (p. 150) from all parts of the body, internal and external, and responds by delivering the appropriate commands to the various organs and muscles (p. 143) to react accordingly.

peripheral nervous system that part of the nervous system (↑) excluding the CNS (↑). It consists of a network of nerves (↓) running through the body of the organism and connected with the CNS.

neurone (*n*) one of the many specially modified cells which make up the nervous system (↑). Each neurone is connected via synapses (p. 151) to others by a single thread-like axon (↓) or nerve fibre and numerous dendrons (↓) which transmit nerve impulses (p. 150) from neurone to neurone .

nerve cell = neurone (↑).

cell body the part of the neurone (↑) with the nucleus (p. 13). Also known as **centron**.

dendron (*n*) a branching process of cytoplasm (p. 10) from the body of a neurone (↑) which ends in a synapse (p.151). They may branch into dendrites .

Nissl's granules granules found in the cytoplasm (p. 10) of a neurone (↑). They are rich in RNA (p. 24).

axon (*n*) the long process of a neurone (↑) filled with axoplasm which normally conducts nerve impulses (p. 150) away from the cell body (↑). The axon is enclosed in a myelin (p. 150) sheath which is bounded by the thin membrane (p. 14), the *neurilemma*, of the *Schwann cell* .

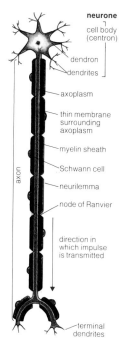

neurone

cell body (centron)

dendron

dendrites

axoplasm

thin membrane surrounding axoplasm

myelin sheath

Schwann cell

axon

neurilemma

node of Ranvier

direction in which impulse is transmitted

terminal dendrites

myelin (*n*) a fatty substance which insulates the
axon (p. 149) and speeds up the transmission
of nerve impulses (↓). In vertebrates (p. 74) not
all axons are myelinated. The myelin sheath is
broken at intervals by constrictions called
nodes of Ranvier.

neuroglia (*n*) specialized cells which protect and
support the central nervous system (p. 149).

nerve impulse one of an interspaced succession
of impulses or signals that are carried between
the neurones (p. 149) via the exchange of
sodium ions and changes in the electrical state
of the neurone. The impulses travel at a
constant speed throughout the nervous system
(p. 149) and the energy for the impulse is not
provided by the stimulus itself.

**transmission of nerve impulse
along nerve**

direction of impulse

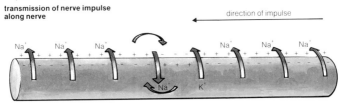

resting potential

membrane polarized: inside
negative, outside positive.
Sodium ions expelled by
sodium pump mechanism

action potential

membrane depolarized:
sodium ions enter axon;
inside positive, outside
negative

resting potential

membrane repolarized

resting potential the state which occurs when a
neurone (p. 149) is inactive so that the neurone
carries a greater negative charge within the
cell and a greater positive charge outside.

action potential the state in which an electrical
charge moves along the membrane (p. 14) of
the axon (p. 149).

sodium pump mechanism the mechanism by
which sodium ions are pumped out of a neurone
(p. 149) as soon as the nerve impulse (↑) has
passed.

polarization (*n*) the process in which sodium ions
are pumped out of the neurone (p. 149) by the
sodium pump mechanism (↑) so that the inside of
the cell is restored to its resting potential (↑).

all or nothing law

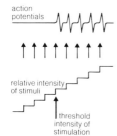

action
potentials

relative intensity
of stimuli

threshold
intensity of
stimulation

synapse

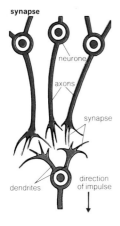

neurone

axons

synapse

dendrites

direction
of impulse

depolarization (n) the process in which the
membrane (p. 14) of the neurone (p. 149)
becomes permeable to the passage of sodium
ions which then enter the cell so that the
cell becomes positively charged.

stimulus (n) any change in the external
environment (p. 218) or the internal state of an
organism which, (via the nervous system
(p. 149) in animals), provokes a response to
that change without supplying the energy for it.

threshold intensity the level of stimulus below
which there is no nervous (p. 149) response of
the stimulated organism.

'all or nothing law' the law which states that an
organism will respond to a stimulus in only two
ways: that is either no nervous (p. 149)
response at all or a response which is of a
degree of intensity which does not vary with the
intensity of the stimulus.

refractory period the length of time which passes
between the passage of a nervous impulse (↑)
through a neurone (p. 149) and its return to the
resting potential (↑). During this period the
neurone cannot further be stimulated.

absolute refractory period a refractory period (↑) in
which a further stimulus of any intensity will result
in the passage of no further nerve impulse (↑).

relative refractory period a refractory period (↑) in
which another, unusually intense stimulus will
result in the passage of a further nerve impulse (↑).

transmission speed the speed at which a nervous
impulse (↑) travels and which is dependent
upon the diameter of the neurone (p. 149).

synapse (n) the gap which exists between
neurones (p. 149) and which is bridged during
the action potential (↑) by a substance secreted
(p. 106) by the neurone.

synaptic transmission the process by which
nervous impulses (↑) are transmitted between
neurones (p. 149) via the synaptic knob (p. 152).
The action potential (↑) stops at the synapse (↑)
but it causes a substance to be released which
travels across the synapse and generates a
new action potential in the neighbouring
neurone.

synaptic knob (*n*) the knob-like ending of the axon (p. 149) which projects into the synapse (p. 151).

acetylcholine (*n*) one of the substances that are released as the action potential (p. 150) in a neurone (p. 149) arrives at the synapse (p. 151). It is specifically produced between a neurone and a muscle (p. 143) cell. There is a special enzyme (p. 28) called acetylcholine esterase which breaks it down so its effect does not continue.

atropine (*n*) a substance which is found in the plant, deadly nightshade, and which acts as a poison by preventing nerve impulses (p. 150) being transmitted from the neurone (p. 149) to the body tissues (p. 83).

strychnine (*n*) a substance which is obtained from the seed of an east Indian tree and which has a powerful stimulating effect on the central nervous system (p. 149), so much so, that in greater than minute quantities it acts as a poison.

adrenaline (*n*) a substance, similar to noradrenaline (↓), released by the adrenal glands (p. 130), which increases the metabolic rate (p. 32) and other functions when it is released into the bloodstream (p. 90) during stress or in preparation for action.

noradrenaline (*n*) one of the substances which is released as the action potential (p. 150) in a neurone (p. 149) arrives at the synapse (p. 151). It is produced in the autonomic nervous system (p. 155). It is also secreted (p. 106) by the adrenal glands (p. 130) and affects cardiac muscle (p. 143) and involuntary muscle (p. 143) etc.

summation (*n*) the process in which the additive effect of nerve impulses (p. 150) arriving at different neurones (p. 149) stimulates the impulse in another neurone while the arrival of just one of the impulses produces no effect.

facilitation (*n*) the process in which the stimulation of a neurone (p. 149) is increased by summation (↑).

reflex action the fundamental and innate (p. 164) response of an animal to a stimulus. For example, the automatic escape reaction away from a source of threat or pain, such as withdrawing the hand from a hot object.

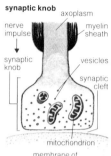

synaptic knob

axoplasm
nerve impulse
myelin sheath
synaptic knob
vesicles
synaptic cleft
mitochondrion
membrane of post-synaptic dendrite

simple reflex arc

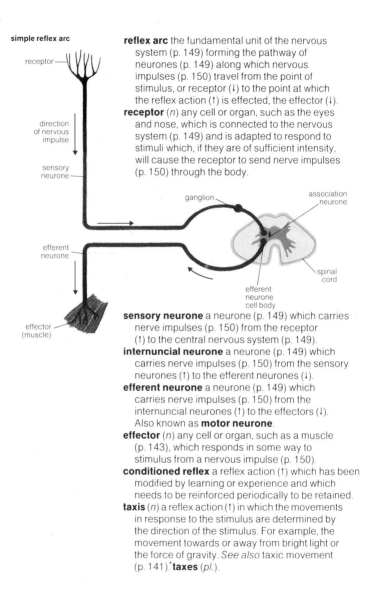

receptor

direction
of nervous
impulse

sensory
neurone

ganglion

association
neurone

efferent
neurone

spinal
cord

efferent
neurone
cell body

effector
(muscle)

reflex arc the fundamental unit of the nervous system (p. 149) forming the pathway of neurones (p. 149) along which nervous impulses (p. 150) travel from the point of stimulus, or receptor (↓) to the point at which the reflex action (↑) is effected, the effector (↓).

receptor (*n*) any cell or organ, such as the eyes and nose, which is connected to the nervous system (p. 149) and is adapted to respond to stimuli which, if they are of sufficient intensity, will cause the receptor to send nerve impulses (p. 150) through the body.

sensory neurone a neurone (p. 149) which carries nerve impulses (p. 150) from the receptor (↑) to the central nervous system (p. 149).

internuncial neurone a neurone (p. 149) which carries nerve impulses (p. 150) from the sensory neurones (↑) to the efferent neurones (↓).

efferent neurone a neurone (p. 149) which carries nerve impulses (p. 150) from the internuncial neurones (↑) to the effectors (↓). Also known as **motor neurone**.

effector (*n*) any cell or organ, such as a muscle (p. 143), which responds in some way to stimulus from a nervous impulse (p. 150).

conditioned reflex a reflex action (↑) which has been modified by learning or experience and which needs to be reinforced periodically to be retained.

taxis (*n*) a reflex action (↑) in which the movements in response to the stimulus are determined by the direction of the stimulus. For example, the movement towards or away from bright light or the force of gravity. *See also* taxic movement (p. 141). **taxes** (*pl.*).

kinesis (*n*) a reflex action (p. 152) in which the rate of movement is affected by the intensity of the stimulus and which is unaffected by its direction. For example, woodlice move faster in drier surroundings than in damp ones.

spinal cord the part of the central nervous system (p. 149) in vertebrates (p. 74) which is contained within a hollow tube running the length of the vertebral column (p. 74) and runs from the medulla oblongata (p. 156). It consists of neurones (p. 149) and nerve fibres with a central canal containing fluid. Pairs of spinal nerves (↓) leave the spinal cord to pass into the body. The spinal cord carries nerve impulses (p. 150) to and from the brain (↓) and the body.

spinal cord

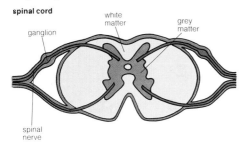

white matter

grey matter

ganglion

spinal nerve

meninges (*n.pl.*) the three membranes (p. 14) which protect the central nervous system (p. 149) of vertebrate (p. 74) animals.

pia mater one of the meninges (↑). The soft, delicate inner membrane (p. 14) next to the central nervous system (p. 149) which is densely packed with blood vessels (p. 127).

arachnoid mater the central of the three meninges (↑) and separated from the pia mater (↑) by fluid-filled spaces.

dura mater one of the meninges (↑). The stiff, tough outer membrane which is in direct contact with the arachnoid mater (↑) and which contains blood vessels (p. 127).

spinal nerve one of the pairs of nerves (p. 149) which arise from the spinal cord (↑) in segments.

meninges

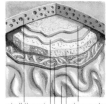

skull (bone)
arachnoid mater
pia mater

dura mater
cerebro-spinal fluid
brain

grey matter nervous tissue (p. 91) that is grey in colour and is found in the centre of the spinal cord (↑) as well as in parts of the brain (↓). It contains large numbers of synapses (p. 151) and consists mainly of nerve (p. 149) cell bodies (p. 149).

white matter nervous tissue (p. 91) that is whitish in colour and is found on the outer region of the spinal cord (↑) as well as in parts of the brain (↓). It connects different parts of the central nervous system (p.149) and consists mainly of axons (p.149).

autonomic nervous system the part of the central nervous system (p. 149) in vertebrates (p. 74) which carries nerve impulses (p. 150) from receptors (p. 153) to the smooth muscle fibres (p. 144) of the heart (p. 124), gut (p. 98) and other internal organs.

sympathetic nervous system the part of the autonomic nervous system (↑) which increases the heart (p. 124) rate and breathing (p. 112) rate, the secretion (p. 106) of adrenaline (p. 152), the blood (p. 90) pressure, and slows the digestion (p. 98) so that the vertebrate's (p. 74) body is prepared for emergency action in response to stimuli.

parasympathetic nervous system the part of the autonomic nervous system (↑) which effectively works in opposition to the sympathetic nervous system (↑) slowing down the heart (p. 124) beat etc. Both systems act in co-ordination to control the rates of action.

ganglion (*n*) a bundle of nerve (p. 149) cell bodies (p. 149) contained within a sheath which, in invertebrates (p. 75) may form part of the central nervous system (p. 149), and in vertebrates (p. 74) are generally found outside the central nervous system. Also, in the brain (↓) some of the masses of grey matter (↑) are referred to as ganglia (*pl.*).

nerve net an interconnecting network of nerve (p.149) cells found in the bodies of some invertebrates (p. 75) to form a simple nervous system (p. 149).

brain (*n*) the part of the central nervous system (p. 149) which effectively co-ordinates the reactions of the whole body of the organism. It forms as an enlargement of the spinal cord (↑) and is situated at the anterior end of the body.

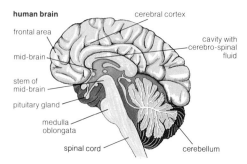

human brain

frontal area

mid-brain

stem of
mid-brain

pituitary gland

medulla
oblongata

spinal cord

cerebral cortex

cavity with
cerebro-spinal
fluid

cerebellum

cerebral hemispheres the paired masses of grey
matter (p. 155), beneath which is white matter
(p. 155), that occur at the front end of the
forepart of the brain (p. 155) and by which many
of the animal's activities are controlled. Each
hemisphere controls actions on the opposite
side of the body from which it is situated.

cerebral cortex the highly convoluted grey matter
(p. 155) that forms part of the cerebral
hemispheres (↑).

corpus callosum the band of nerve (p. 149)
fibres which connects the cerebral hemispheres
(↑) allowing their action to be co-ordinated.

medulla oblongata the continuation of the spinal
cord (p. 154) with the hind region of the brain
(p. 155). It contains centres of grey matter
(p. 155) which are responsible for controlling
many of the major functions and reflexes
(p. 153) of the body, for example, the medulla
oblongata contains the respiratory centre (p.117).

cerebellum (*n*) the part of the brain (p. 155) lying
between the medulla oblongata (↑) and the
cerebral hemispheres (↑) which is deeply
convoluted and is responsible for controlling
voluntary muscle (p. 143) action which is
stimulated by the cerebral hemispheres.

hypothalamus (*n*) that region of the forepart of
the brain (p. 155) which is responsible for
monitoring and regulating metabolic (p. 26)
functions such as body temperature, eating,
drinking and excretion (p. 134). It controls the
activity of the pituitary gland (↓).

thalamus (*n*) the part of the brain (p. 155) which carries and co-ordinates nerve impulses (p. 150) from the cerebral hemispheres (↑).

pituitary gland a gland (p. 87) in the brain (p. 155) which secretes (p. 106) a number of hormones (p. 130) that in turn stimulate the secretion of hormones from other glands to affect such metabolic (p. 26) processes as growth, secretion of adrenaline (p. 152), milk production, and so on.

pineal body a gland (p. 87) found as an outgrowth on the top of the brain (p. 155) and which may be responsible for secreting (p. 106) a hormone (p. 130) associated with colour change.

human ear

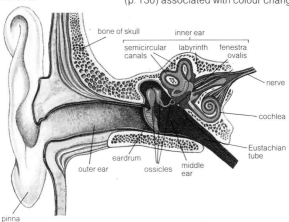

bone of skull

inner ear

semicircular canals   labyrinth   fenestra ovalis

nerve

cochlea

Eustachian tube

eardrum   ossicles   middle ear

outer ear

pinna

ear (*n*) one of the pair of sense organs, situated on either side of the head in vertebrates (p. 74), which are used for hearing (p. 159) and balance (p. 159).

outer ear the tube which leads from the outside of the head to the eardrum (p. 158). In amphibians (p. 77) and some reptiles (p. 78), it is not present because the eardrum is situated at the skin surface.

pinna (*n*) the part of the outer ear (↑), present in mammals (p. 80), situated on the outside of the head and consisting of a flap of skin and cartilage (p. 90), which helps to direct sound into the ear (↑).

inner ear the innermost part of the ear (p. 157,)
which is situated within the skull and which
detects sound as well as the position of the
animal in relation to gravity and acceleration, so
enabling the animal to balance. It is fluid filled
and is connected to the brain (p. 155) by an
auditory nerve (p. 149) so that it is able to convert
sound waves into nervous impulses (p. 150). It
is made up of a labyrinth of membraneous
(p. 14) tubes contained within bony cavities.

middle ear an air-filled cavity situated between
the outer ear (p. 157) and the inner ear (↑). It is
separated from the outer ear by the eardrum (↓)
and, in mammals (p. 80), contains three small
bones or ossicles (↓).

eardrum (*n*) a thin, double membrane (p. 14) of
epidermis (p. 131) separating the outer ear
(p.157) from the middle ear (↑) and which is
caused to vibrate by sound waves. These
vibrations are then transmitted through the
middle ear, where their force is amplified, into
the inner ear (↑). Also known as **tympanic
membrane**.

fenestra ovalis a small, oval, membraneous
(p. 14) window which connects the middle ear
(↑) with the inner ear (↑) allowing vibrations from
the eardrum (↑) to be transmitted to the inner
ear. It is twenty times smaller than the eardrum
so that the force of the vibrations is increased.

fenestra rotunda a small, round, membraneous
(p. 14) window which connects the inner ear (↑)
with the middle ear (↑) and which bulges into
the middle ear as vibrations of the fenestra ovalis
(↑) cause pressure increases in the inner ear.

ear ossicle one of usually three small bones that
occur in the middle ear (↑) of mammals (p. 80)
and which, by acting as levers, transmit and
increase the force of vibrations produced by the
eardrum (↑) and carry them to the inner ear (↑).

malleus (*n*) a hammer-shaped ear ossicle (↑)
which is connected with the eardrum (↑).

incus (*n*) an anvil-shaped ear ossicle (↑) situated
between the malleus (↑) and the stapes (↓).

stapes (*n*) a stirrup-shaped ear ossicle (↑)
which is connected with the fenestra ovalis (↑).

ear ossicles

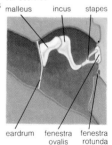

malleus   incus   stapes

eardrum   fenestra   fenestra
ovalis   rotunda

section through cochlea

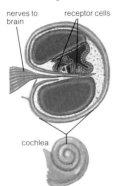

nerves to brain

receptor cells

cochlea

Eustachian tube a tube which connects the middle ear (↑) with the back of the throat. It is normally closed but opens during yawning and swallowing to balance the pressure on either side of the eardrum (↑) thus preventing the eardrum from bursting.

vestibular apparatus the apparatus contained in a cavity in the inner ear (↑) immediately above and behind the fenestra ovalis (↑) and which contains the organs concerned with the sense of balance and posture.

cochlea (n) a spirally coiled tube, which is a projection of the saccule (p. 160), and found within the inner ear (↑). It is concerned with sensing the pitch (↓) of the sound waves entering the ear (p. 157).

hearing (n) the sense whereby sound waves or vibrations enter the outer ear (p. 157) and cause the eardrum (↑) to vibrate. In turn, these vibrations are transmitted through the middle ear (↑) and into the inner ear (↑) where they are converted into nervous impulses (p. 150) and transmitted to the brain (p. 155).

intensity (n) the degree of loudness or softness of sound. If a sound entering the ear (p. 157) is very loud, muscles (p. 143) attached to the ear ossicles (↑) prevent them from vibrating too much.

pitch (n) the degree of height or depth of a sound which depends on the frequency of the sound waves – those of a high frequency are referred to as high and vice versa. Different parts of the cochlea (↑) respond to sounds of different pitch.

semicircular canals

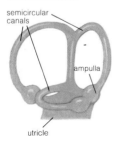

semicircular canals

ampulla

utricle

balance (n) the ability of the animal to orient itself properly in relation to the force of gravity. Animals rely on information received by apparatus within the inner ear (↑) and by the eyes and other senses to achieve balance and posture.

semicircular canal one of three looped tubes positioned at right angles to one another within the inner ear (↑). They contain fluid which flows in response to movement of the head. The movement of the fluid is detected by the sensory hairs in ampullae (p. 160) at the ends of the canals.

ampulla (*n*) a swelling at the end of each semicircular canal (p. 159). It bears a gelatinous cupula (↓), sensory hairs, and receptor (p. 153) cells which are responsible for transmitting information to the brain (p. 155) via the auditory nerve (p. 149). **ampullae** (*pl.*).

utricle (*n*) a fluid-filled sac within the inner ear (p. 158) from which arise the semicircular canals (p. 159). Within the fluid of the utricle are otoliths (↓) of calcium carbonate. If the head is tilted, the otoliths are pulled downwards by gravity and in turn pull on sensory fibres attached to the wall of the utricle.

saccule (*n*) a lower, fluid-filled cavity in the inner ear (p. 158) from which arises the cochlea (p. 159). Like the utricle (↑), it also contains otoliths (↓) which respond to the orientation of the head with respect to the force of gravity.

cupula (*n*) the gelatinous body which forms part of the ampulla (↑) and which is displaced by fluid that moves in response to the position of the head.

otolith (*n*) one of the granules of calcium carbonate present in the fluid of the utricle (↑) and saccule (↑) which respond to tilting movements of the head by force of gravity.

eye (*n*) the sense organ of sight which is sensitive to the direction and intensity of light and which, in vertebrates (p. 74), is also able to form complex images of the outside world which are transmitted to the brain (p. 155) via the optic nerve (p. 149). The eyes of most animals are roughly spherical in shape, and in vertebrates are contained in depressions within the skull, to which they are attached by muscles (p. 143).

retina (*n*) the light-sensitive inner layer of the eye (↑) which contains rod-shaped receptor (p. 153) cells and cone-shaped receptor cells. Nerve (p. 149) fibres leave the retina and join together to form the optic nerve.

choroid layer a layer of tissue surrounding the eye (↑) between the retina (↑) and the sclerotic layer (↓). It contains pigments (p. 126) to reduce reflections within the eye and blood vessels (p. 127) which supply oxygen to the eye.

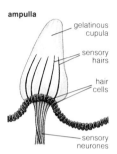

ampulla

gelatinous cupula

sensory hairs

hair cells

sensory neurones

human eye

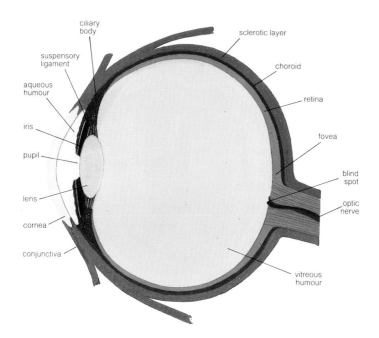

ciliary body

sclerotic layer

suspensory ligament

choroid

aqueous humour

retina

iris

fovea

pupil

blind spot

optic nerve

lens

cornea

conjunctiva

vitreous humour

sclerotic layer the tough, fibrous (p. 143) non-elastic layer which surrounds and protects the eye (↑) and is continuous with the cornea (↓).

cornea (*n*) the disc-shaped area at the front of the eye (↑) which is continuous with the sclerotic layer (↑) and which is transparent to light. It is curved so that light passing through it is refracted and begins to converge before reaching the lens (p. 162). Indeed, in land-living mammals (p. 80) this is the main element of the eye's focusing power.

lens accommodation

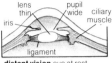

distant vision eye at rest

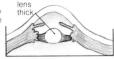

close vision

lens (*n*) a transparent disc which is convex on both faces and which is attached to the ciliary body (↓) by suspensory ligaments (↓). It consists of an elastic, jelly-like material held in a skin. When the ciliary muscles (↓) contract, the convexity of the lens is increased so that light rays entering the eye can be focused on the retina (p. 160). This allows for the fine focusing power of the eye (p. 160).

refraction (*n*) the change in direction of light which takes place as it crosses a boundary between two substances.

convex (*adj*) of a lens which causes light passing through it to move closer together.

concave (*adj*) of a lens which causes light passing through it to move further apart rather than closer together. *See also* convex (↑).

suspensory ligament the fibrous (p. 143) ligament (p. 146) which holds the lens (↑) in position.

ciliary body the circular, thickened outer edge of the choroid (p. 160) at the front of the eye which contains the ciliary muscles and to which the lens (↑) is attached. It also contains glands (p. 87) which secrete (p. 106) the aqueous humour (↓).

ciliary muscles *see* ciliary body (↑).

iris (*n*) a ring of opaque tissue (p. 83) that is continuous with the choroid (p. 160) and which has a hole or pupil (↓) in the centre through which light can pass. There are circular muscles (p.143) and radial muscles surrounding the pupil which increase or decrease the size of the pupil in accordance with the intensity of light.

pupil (*n*) the hole in the centre of the iris (↑) through which light enters the eye (p. 160). It is usually circular but may be other shapes in some animals.

refraction

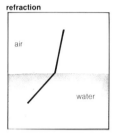

convex lens

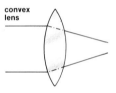

concave lens

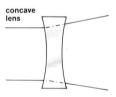

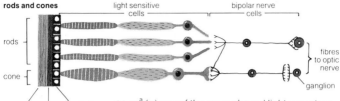

rods and cones

light sensitive cells

bipolar nerve cells

rods

cone

fibres to optic nerve

ganglion

sclerotic layer choroid epithelium

normal sight

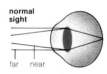

far near

long sight (hypermetropia)
eyeball shorter than normal

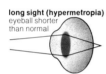

convex lens

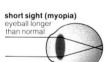

short sight (myopia)
eyeball longer than normal

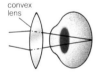

concave lens

cone[a] (n) one of the cone-shaped light receptors (p. 153) present in the retina (p. 160). Cones contain three different pigments (p. 126) which are in turn sensitive to red, green, and blue light so that cones are primarily responsible for colour vision. Cones are concentrated mainly in and around the fovea (↓) and are not present at the edge of the retina. They are also receptors of light of high intensity.

fovea (n) a small, central depression in the retina (p. 160) in which most of the receptor (p. 153) cells, especially the cones (↑) are concentrated. It is directly opposite the lens and provides the main area for acute and accurate daylight vision (↓).

daylight vision vision of great sharpness which takes place in bright light since most of the light entering the eye (p. 160) falls on the fovea (↑).

rod (n) one of the rod-shaped light receptors (p. 153) present in the retina (p. 160) which are much more sensitive to light of low intensity but are not sensitive to colour. They are not present in the fovea (↑) and increase in numbers towards the edges of the retina. They are also sensitive to movements.

night vision vision which takes place in light of low intensity using the rods (↑).

aqueous humour a watery fluid that fills the space between the cornea (p. 161) and the vitreous humour (↓) and in which lie the lens (↑) and the iris (↑). It is secreted (p. 87) by glands (p. 106) in the ciliary body (↑).

vitreous humour the jelly-like fluid that fills the space behind the lens (↑).

blind-spot the area of the retina (p. 160) from which the optic nerve (p. 149) leaves the eye. It contains no rods (↑) or cones (↑) so that no image is recorded on this part of the retina.

behaviour (*n*) all observable activities carried out by an animal in response to its external and internal environment (p. 218).

ethology (*n*) the study or science of the behaviour (↑) of animals in their natural environment (p. 218).

instinctive behaviour behaviour (↑) which is believed to be controlled by the genes (p. 196) and which is unaffected by experience, e.g. the courtship behaviour of many animals, such as fish and birds, is stimulated by a particular signal provided the animal is sexually mature and has the appropriate level of sex hormones (p. 130) in its body.

innate behaviour behaviour (↑) that does not have to be learned. *See also* instinctive behaviour (↑).

learned behaviour behaviour (↑) in which the response to stimuli is affected by the experience of the individual animal to gain the best advantage from a situation.

habituation (*n*) a learned behaviour (↑) in which the response to a stimulus is reduced by the constant repetition of the stimulus.

associative learning a learned behaviour (↑) in which the animal learns to associate a stimulus with another that normally produces a reflex action (p. 152). For example, dogs respond to the sight of food by salivating. Pavlov's dogs learned to associate the sight of food with the ringing of a bell and would then salivate on hearing the bell without seeing the food.

imprinting (*n*) a learned behaviour (↑) which takes place during the very early stages of an animal's life so that e.g. the animal continues to follow the first object on which its attention is fixed by sight, sound, smell or touch. This is usually the animal's parent.

exploration (*n*) the process through which animals learn about their environment (p. 218) while they are young by play and contact with other animals.

orientation (*n*) the reflex action (p. 152) in which the animal changes the position of part or the whole of its body in response to a stimulus. For example, an animal might turn its head or prick up its ears in response to a sudden or unusual sound.

releaser (*n*) a stimulus which releases instinctive behaviour (↑) in an animal.

incomplete metamorphosis
e.g. locust

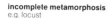

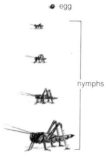

egg

nymphs

adult/imago

complete metamorphosis
e.g. butterfly

egg

larva

pupa

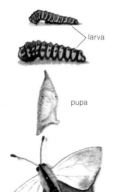

adult/imago

growth (*n*) the permanent increase in size and dry mass of an organism which takes place as the cells absorb (p. 81) materials, expand, and then divide. The temporary take-up of water cannot be considered as growth.

growth rate the amount of growth which takes place in a given unit of time.

incomplete metamorphosis the change which takes place from young to adult form in which the young closely resembles the adult.

instar (*n*) a stage through which an insect (p. 69) passes during incomplete metamorphosis (↑).

ecdysis (*n*) the periodic moulting or shedding of the external cuticle (p. 145) allowing growth, which takes place between instars (↑) during incomplete metamorphosis (↑).

nymph (*n*) the early instar (↑) or young of an insect (p. 69) which is small, sexually immature, and unable to fly.

complete metamorphosis the change which takes place from young to adult form in which the young do not resemble the adult and which can occur through a pupal (↓) stage.

larva (*n*) the immature stage in the life cycle of an animal, for example, an insect (p. 69) which undergoes metamorphosis (p. 70). The larva is usually different in structure and appearance from the adult. It hatches from the egg and is able to fend for itself. **larvae** (*pl.*), **larval** (*adj*).

pupa (*n*) the stage between larva (↑) and adult in an insect (p. 69) in which movement and feeding stop and metamorphosis (p. 70) takes place. **pupae** (*pl.*), **pupate** (*v*).

imago (*n*) the sexually mature, adult stage of an insect's (p. 69) development.

corpora allata a pair of glands (p. 87) in the head of an insect (p. 69) which secrete (p. 106) a hormone (p. 130) that encourages the growth of larval (↑) structures and discourages that of adult structures.

ecdysial glands a pair of glands (p. 87) in the head of an insect (p. 69) which secrete (p. 106) a hormone (p. 130) that stimulates ecdysis (↑) and growth.

morphogenesis (*n*) the process in which the overall form of the organs of an organism is developed, leading to the development of the whole organism.

differentiation (*n*) the process in which cells (unspecialized) change in form and function, during the development of the organism, to give all of the different types of specialized cells that characterize that organism.

neotony (*n*) the retention in some animals of larval (p. 165) or embryonic (↓) features, either temporarily or permanently beyond the stage at which they would normally be lost. It is thought to be important in evolutionary (p. 208) development. For example, humans retain certain resemblances to young apes.

embryology (*n*) the study of embryos (↓).

embryo (*n*) the stage in the development of an organism between the zygote (↓) and hatching, birth, or germination (p. 168). **embryonic** (*adj*).

zygote (*n*) the diploid (p. 36) cell which results from the fusion of a haploid (p. 36) male gamete (p. 175) or spermatozoon (p. 188) and a haploid female gamete or ovum (p. 178 and p. 190).

cleavage (*n*) the process in which the nuclei (p. 13) and cytoplasm (p. 10) of the fertilized (p. 175) zygote (↑) divide mitotically (p. 37) to form separate cells. Also known as **segmentation**.

blastula (*n*) the embryonic (↑) structure or mass of small cells which results from cleavage (↑).

blastocoel (*n*) the cavity which occurs in the centre of a blastula (↑) during the final stages of cleavage (↑).

gastrulation (*n*) the process which follows cleavage (↑) in which cell movements occur to form a gastrula (↓) that will eventually lead to the formation of the main organs of the animal. In simple instances, part of the blastula (↑) wall folds inwards to form a hollow gastrula.

gastrula (*n*) the stage in the development of an animal embryo (↑) which comprises a two-layered wall of cells surrounding a cavity known as the archenteron.

ectoderm (*n*) the external germ layer (↓) of an embryo (↑) which develops into hair, various glands (p. 87), CNS, the lining of the mouth etc.

endoderm (*n*) the internal germ layer (↓) of the embryo (↑) which develops into the lining of the gut (p. 98) and its associated organs.

stages of cleavage in *Amphioxus*

single cell zygote

2-cell stage

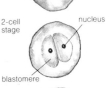

nucleus

blastomere

8-cell stage

blastula

blastula in section

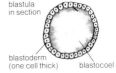

blastoderm (one cell thick)   blastocoel

gastrulation

1.

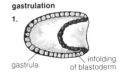

gastrula

infolding of blastoderm

2.

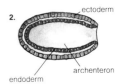

ectoderm

endoderm

archenteron

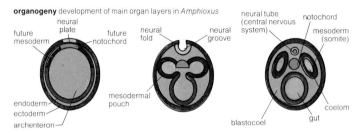

organogeny development of main organ layers in *Amphioxus*

organogeny (*n*) the formation of the organs during growth. **organogenesis** (*n*).

notochord (*n*) the flexible skeletal (p. 145) rod which is present at some stage in the development of all chordates (p. 74). It stretches from the central nervous system (p. 149) to the gut (p. 98) and, in vertebrates (p. 74), it persists as remnants in the backbone throughout the life of the animal although it is present primarily during the development of the embryo (↑).

neural tube the part of the brain (p. 155) and vertebral column (p. 74) which forms first during the growth of the embryo (↑).

mesoderm (*n*) the germ layer (↓) lying between the ectoderm (↑) and endoderm (↑) and which gives rise to connective tissue (p. 88), blood (p. 90), and muscles (p. 143), etc.

germ layer one of the two or three main layers of cells which can be seen in an embryo (↑) after gastrulation (↑). The endoderm (↑), ectoderm (↑) and mesoderm (↑) are all germ layers.

coelom (*n*) a fluid-filled cavity in the mesoderm (↑) of triploblastic (p. 62) animals which, in higher animals, forms the main body cavity in which the gut (p. 98) and other organs are suspended so that their muscular (p. 143) contractions may be independent of those of the body wall.

somite (*n*) any one of the blocks of mesoderm (↑) tissue (p. 83) that flank the notochord (↑) as parallel strips and which develop into blocks of muscle (p. 143), parts of the kidneys (p. 136) and parts of the axial skeleton (p. 145).

myotome (*n*) the part of a somite (↑) that develops into striped muscle (p. 143) tissue (p. 83).

germination (*n*) the first outward sign of the growth of the seeds or spores (p. 178) of a plant which takes place when the conditions of moisture, temperature, light, and oxygen are suitable. **germinate** (*v*).

hydration phase the stage of germination (↑) in which the seed absorbs (p. 81) water and the activity of the cytoplasm (p. 10) begins.

metabolic phase the stage of germination (↑) in which, under enzyme (p. 28) control, the water absorbed (p. 81) during the hydration phase (↑) hydrolyzes (p. 16) the stored food materials into the materials needed for growth.

plumule (*n*) the first apical (↓) leaves and stem which form part of the embryonic (p. 166) shoot of a spermatophyte (p. 57).

radicle (*n*) the first root in an embryonic (p. 166) spermatophyte (p. 57) which later develops into the plant's rooting system.

cotyledon (*n*) the first, simple, leaf-like structure that forms within a seed. Plants, e.g. grasses and cereals, have only one and are called monocotyledons (p. 58) while other flowering plants have two and are called dicotyledons (p.57). Cotyledons contain no chlorophyll (p.12) at first and may function as food reserves for the germinating (↑) plant but, in most dicotyledons, the cotyledons emerge above ground, turn green and photosynthesize (p. 93).

endosperm (*n*) the layer of tissue (p. 83) which surrounds the embryo (p. 166) in some spermatophytes (p. 57). It supplies nourishment to the developing embryo but, in some plants e.g. peas and beans, it has been absorbed (p.81) by the cotyledons (↑) by the time the seed is fully developed while, in others such as wheat, it is not absorbed until the seed germinates (↑).

testa (*n*) a hard, tough, protective coat which surrounds the seed and shields it from mechanical damage or the invasion of fungi (p. 46) and bacteria (p. 42). Also known as **seed coat**.

epicotyl (*n*) the part of the plumule (↑) which lies above the attachment point of the cotyledons (↑).

hypocotyl (*n*) the part of the plumule (↑) which lies below the attachment point of the cotyledons (↑).

types of seed

cotyledonous
food stored in cotyledons
e.g. bean

testa
micropyle
hilum

embryo

plumule
epicotyl
hypocotyl
radicle

cotyledons

endospermic
most food stored in endosperm
e.g. maize

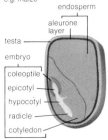

endosperm
aleurone layer

testa

embryo

coleoptile
epicotyl
hypocotyl
radicle
cotyledon

hypogeal germination germination (↑) in which the cotyledons (↑) remain below the surface of the soil, e.g. in broad beans.

epigeal germination germination (↑) in which the cotyledons (↑) emerge above the soil surface and form the first photosynthetic (p. 93) seed leaves e.g. in lettuces.

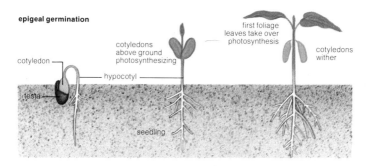

epigeal germination

first foliage leaves take over photosynthesis

cotyledons above ground photosynthesizing

cotyledon

hypocotyl

testa

seedling

cotyledons wither

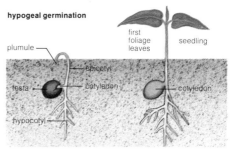

hypogeal germination

first foliage leaves

seedling

plumule

epicotyl

testa

cotyledon

cotyledon

hypocotyl

meristem (*n*) the part of an actively growing plant where cells are dividing and new permanent plant tissue (p. 83) is formed.

apical meristem a meristem (↑) which occurs at shoot and root tips. The division of these cells which are small and contain granular cytoplasm (p. 10) with small vacuoles (p. 11), results in the growth of stems and roots.

apex (*n*) the top or pointed end of an object.
apical (*adj*).

lateral meristem a meristem (p. 169), including
the vascular cambium (p. 86) and phellogen
(p. 172), which occurs along the roots and
stems of dicotyledons (p. 57) plants and
which is made up of long, thin cells that give
rise to xylem (p. 84) and phloem (p. 84).

lateral (*adj*) on, at or about the side of something.

ground meristem the part of the apical meristem
(p. 169) from which pith (p. 86), cortex (p. 86),
medullary rays (p. 86) and mesophyll (p. 86)
are formed.

tunica (*n*) one of the two layers of tissue (p. 83)
comprising the apical meristem (p. 169). It is
the outer layer of tissue and may itself be made
up of one or more layers of cells in which the
division takes place at right angles to the
surface of the plant (anticlinally).

corpus (*n*) one of the two layers of tissue (p. 83)
comprising the apical meristem (p. 169). It is
the inner layer of tissue and the division of cells
occurs irregularly.

zone of cell division the part of the apex (p. 169)
of a root or shoot which includes the apical
meristem (p. 169) and the leaf primordium (↓) or
the root cap (p. 81).

zone of expansion the part of the apex (p. 169)
of a root or shoot which lies behind the zone of
cell division (↑) and in which the cells elongate
and expand.

zone of differentiation the part of the apex
(p. 169) of a root or shoot which lies behind the
zone of expansion (↑) and in which the cells
·differentiate into the form and function of parts
of the plant that characterize it.

primordium (*n*) the group of cells in the apex of a
shoot or root which differentiates into a leaf etc.
primordia (*pl.*).

primary growth the growth which takes place
only in the meristems (p. 169) which were
present in the embryo (p. 166). These include
the apical meristems (p. 169) and primary
growth results largely in the increase in length.

secondary growth the growth which takes place
in the lateral meristems (↑) and which results in
increase in girth rather than in length.

**meristems of shoot
and root**

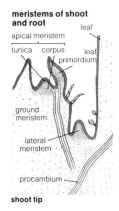

shoot tip

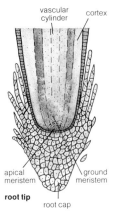

root tip

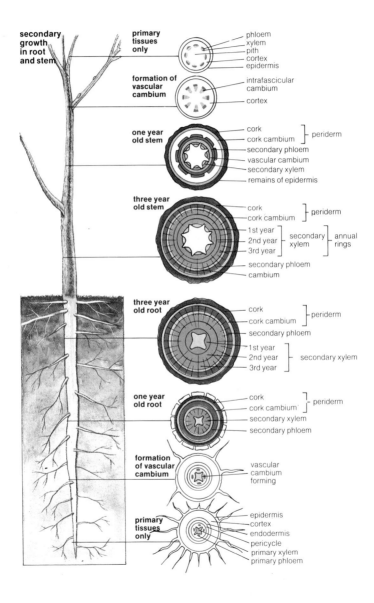

secondary growth in root and stem

primary tissues only
- phloem
- xylem
- pith
- cortex
- epidermis

formation of vascular cambium
- intrafascicular cambium
- cortex

one year old stem
- cork
- cork cambium } periderm
- secondary phloem
- vascular cambium
- secondary xylem
- remains of epidermis

three year old stem
- cork
- cork cambium } periderm
- 1st year
- 2nd year } secondary xylem } annual rings
- 3rd year
- secondary phloem
- cambium

three year old root
- cork
- cork cambium } periderm
- secondary phloem
- 1st year
- 2nd year } secondary xylem
- 3rd year

one year old root
- cork
- cork cambium } periderm
- secondary xylem
- secondary phloem

formation of vascular cambium
- vascular cambium forming

primary tissues only
- epidermis
- cortex
- endodermis
- pericycle
- primary xylem
- primary phloem

fascicular (*adj*) of meristematic (p. 169) cambium
    (p. 86) between the xylem (p. 84) and phloem
    (p. 84) in vascular tissue (p. 83).

interfascicular (*adj*) of meristematic (p. 169)
    cambium (p. 86) that consists of a single layer
    of actively dividing cells between the phloem
    (p. 84) and xylem (p. 84) bundles in stems.

secondary xylem xylem (p. 84) that has been
    formed by the vascular cambium (p. 86)
    following the formation of the primary tissues
    (p. 83).

secondary phloem phloem (p. 84) that has been
    produced by the cambium (p. 86) following the
    formation of the primary tissues (p. 83).

annual rings the rings of lighter and darker wood
    that can be seen in a cross-section of the trunk
    of a tree living in temperate conditions. Each
    pair marks the annual increase in the girth of
    the tree as a result of activity of the cambium
    (p. 86). The lighter ring is large-celled xylem
    (p. 84) tissue (p. 83) produced in spring and the
    darker wood is smaller-celled summer wood.

phellogen (*n*) a layer of cells immediately beneath
    the epidermis (p. 131) of the stems undergoing
    secondary growth. It is a lateral meristem
    (p. 170) whose cells give rise to the phellem (↓)
    and phelloderm (↓). Also known as **cork
    cambium.**

bark (*n*) the protective outer layer of the stems of
    woody plants. It may consist of cork cells only
    or alternating layers of cork (↓) and dead
    phloem (p. 84).

cork (*n*) = phellem (↓).

phellem (*n*) an outer layer of dead, waterproof
    cells formed from the activity of the phellogen
    (↑) on the stems of woody plants.

phelloderm (*n*) the inner layer of the bark (↑)
    produced by the activity of the phellogen (↑).

suberin (*n*) a mixture of substances derived from
    fatty acids (p. 20) and present in the walls of
    phellem (↑) cells rendering these cells waterproof.

exogenous (*adj*) of branching generated on the
    outside of the plant.

endogenous (*adj*) of branching generated on the
    inside of the plant.

annual rings

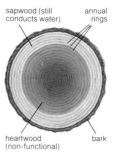

sapwood (still
conducts water)

annual
rings

heartwood
(non-functional)

bark

reproduction (*n*) the means whereby organisms ensure the continued existence of the species (p. 40) beyond the life span of an individual by generating new individuals.

sexual reproduction the generation of new individuals of an organism to continue the life of the species (p. 40) by fusion of haploid (p. 36) nuclei (p. 13) or gametes (p. 175) to form a zygote (p. 166). In most animals a highly motile male spermatozoon (p. 188), generated by the testes (p. 187) and produced in large numbers, unites with a non-motile female ovum (p. 190) produced in small numbers in the ovary (p. 189).

asexual reproduction the generation of new individuals of an organism to continue the life of the species (p. 40) from a single parent by such means as budding (↓) or sporulation (↓). Multiplication is rapid and the offspring are genetically (p. 191) identical to one another and to the parent. For example, binary fission (p. 44) can occur very rapidly and is exponential so that one cell divides into two, two into four, four into eight, and so on, and all the cells are identical to the parent.

motile (*adj*) able to move. **motility** (*n*).

non-motile (*adj*) unable to move.

budding (*n*) asexual reproduction (↑), typical of corals (p. 61) and sponges, in which the parent produces an outgrowth or bud which develops into a new individual.

fragmentation (*n*) asexual reproduction (↑), occurring only in simple organisms such as sponges and green algae (p. 44), in which the parent fragments or breaks up and each fragment develops into a new individual.

sporulation (*n*) asexual reproduction (↑), typical of fungi (p. 46), in which the parent produces often large numbers of small, usually lightweight, single-celled structures, or spores (p. 178), which detach from the parent and may be widely distributed by wind or other mechanisms. Provided they fall into suitable conditions each spore germinates (p. 168) to produce a new individual.

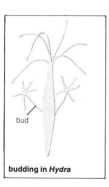

bud

budding in *Hydra*

vegetative propagation asexual reproduction (p. 173), occurring in plants, in which part of the plant, such as a leaf or even a specially produced shoot, detaches from the plant and develops into a new individual.

perennating organ any special structure, found in some biennial (p. 58) and perennial (p. 58) plants, which enables them to survive hostile conditions, such as drought. If more than one structure is produced, asexual reproduction (p. 173) is also achieved. When the conditions deteriorate, the plant dies back, leaving only the perennating organ which, with the onset of suitable conditions, develops into a new individual or individuals.

runner (n) an organ in the form of a stem that develops from an axillary (p. 83) bud, runs along the ground, and produces new individuals at its axillary buds or at the terminal bud only.

stolon (n) an organ which takes the form of a long, erect branch that eventually bends over, under its own weight, so that the tip touches the ground and roots. At the axillary (p. 83) bud a new shoot grows into a new individual.

rhizome (n) a perennating organ (↑) in which the stem of the plant remains below ground and continues to grow horizontally.

bulb (n) a perennating organ (↑) which takes the form of an underground condensed shoot with a short stem and fleshy leaves that are close together, overlap and form the food store for the plant. At each leaf base is a bud which can grow into a new bulb. In the growing season, the plant produces leaves and flowers, and exhausts the food store of the bulb. But the new plant makes more food material by photosynthesis (p. 93) and a new bulb (or bulbs) is formed at the leaf base bud.

corm (n) a perennating organ (↑) which takes the form of a special enlarged, fleshy, underground stem that acts as a food store.

tuber (n) a perennating organ (↑) similar to a rhizome (↑) in which food is stored in the underground stem or root as swellings which eventually detach from the parent plant.

vegetative propagation

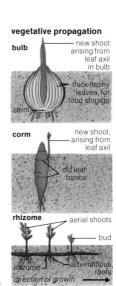

bulb — new shoot arising from leaf axil in bulb — thick fleshy leaves, for food storage — stem

corm — new shoot, arising from leaf axil — old leaf bases

rhizome — aerial shoots — bud — rhizome — adventitious roots — direction of growth →

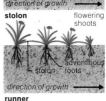

stolon — flowering shoots — stolon — adventitious roots — direction of growth →

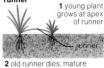

runner — **1** young plant grows at apex of runner — runner

2 old runner dies, mature new plant produces new runner — runner

sexual reproduction

gametes (haploid)

fusion

zygote (diploid)

anisogamy

gametes

fusion

zygote

isogamy

gametes

fusion

zygote

oogamy

motile gamete

gamete

fusion

zygote

apomixis (*n*) asexual reproduction (p. 173) which superficially resembles sexual reproduction (p. 173) although fertilization (↓) and meiosis (p. 38) do not occur. **apomictic** (*adj*).

fertilization (*n*) the process in sexual reproduction (p. 173) in which the nucleus (p. 13) of a haploid (p. 36) male gamete (↓) fuses with the nucleus of a haploid female gamete to form a diploid (p. 36) zygote (p. 166).

fertile (*adj*) of organisms able to produce young.

mature (*adj*) fully grown, fully developed.

immature (*adj*) not mature (↑).

gamete (*n*) a reproductive (p. 173) or sex cell. Each gamete is haploid (p. 36) and male gametes, or spermatozoons (p. 188) which are small and highly motile, fuse with larger female gametes or ova (p. 190) in the process of fertilization (↑) to form diploid (p. 36) zygotes (p. 166) which are capable of developing into new individuals.

gametangium (*n*) an organ which produces gametes (↑). **gametangia** (*pl.*).

syngamy (*n*) the actual fusion of two gametes (↑) which occurs during fertilization (↑).

isogametes (*n.pl.*) gametes (↑), produced by some organisms, e.g. some fungi (p. 46), which are not differentiated into male or female forms. All the gametes produced by the organism are similar.

anisogametes (*n.pl.*) gametes (↑) which are differentiated in some way either simply by size or by size and form.

heterogametes (*n.pl.*) anisogametes (↑) which are differentiated by size and form into small, highly motile male spermatozoons (p. 188), which are produced in large numbers, and large non-motile female ova (p. 190).

oogamy (*n*) fertilization (↑) which takes place by the union of heterogametes (↑).

dioecious (*adj*) of organisms in which the sex organs are borne on separate individuals which are themselves then described as either males or females.

monoecious (*adj*) of organisms in which the male and female sex organs are borne on the same individual.

hermaphrodite (*n, adj*) = monoecious (↑).

parthenogenesis (*n*) reproduction (p. 173), occurring in some plants and animals such as the dandelion or aphids, in which the female gametes (p. 175) develop into new individuals without having been fertilized (p. 175). The offspring of parthenogenesis are always female and usually genetically (p. 196) identical with the parent and with one another. If the ovum (p. 178 and p. 190) has been produced by meiosis (p. 38), the offspring are haploid (p. 36) while, if it has been produced by mitosis (p. 37) the offspring are diploid (p. 36).

alternation of generations a life cycle of an organism in which reproduction (p. 173) alternates with each generation between sexual reproduction (p. 173) and asexual reproduction (p. 173). It is found, for example, among Coelenterata (p. 60) which have both a polyp (p. 61) and a medusa (p. 61) stage and among bryophytes (p. 52) in which haploid (p. 36) gametes (p. 175) from one stage – gametophyte (↓) – fuse to form a diploid (p. 36) zygote (p. 166) which germinates (p. 168) to form a sporophyte (↓) that, in turn, produces haploid spores (p. 178) by meiosis (p. 38). These develop into a haploid plant body (gametophyte). Each of the generations may be quite different in form.

generation (*n*) a set of individuals of the same stage of development or age, or the time taken for one individual to reproduce (p. 173) and for the progeny (p. 200) to develop to the same stage as the parent.

haplontic (*adj*) of a life cycle, found in some algae (p. 44) and fungi (p. 46), in which a haploid (p. 36) adult form occurs by meiosis (p. 38) of the diploid (p. 36) zygote (p. 166).

diplontic (*adj*) of a life cycle, found in all animals, as well as some algae (p. 44) and fungi (p. 46), in which haploid (p. 36) gametes (p. 175) are produced by meiosis (p. 38) from the diploid (p. 36) adults.

diplohaplontic (*adj*) of a life cycle, found in most plants, in which a diploid (p. 36) sporophyte (↓) generation alternates with a haploid (p. 36) gametophyte (↓) generation.

	gametophyte haploid	sporophyte diploid	
bryophytes			sporophyte dependent on gametophyte
pteridophytes		young sporophyte first leaf first root	sporophyte dependent on gametophyte only in very young stage
gymnosperms	pollen grain ♂ ♀ in ovule		gametophyte dependent on sporophyte
angiosperms	pollen grains ♂ ♀ embryo sac in ovule		gametophyte dependent on sporophyte

alternation of generations and the major plant divisions

sporophyte(n) the stage of an alternation of generations (↑), found in most plants, in which the diploid (p. 36) plant produces spores (p. 178) by meiosis (p. 38) which then germinate (p. 168) to produce the gametophyte (↓).

gametophyte (n) the stage of an alternation of generations (↑), found in most plants, in which the haploid (p. 36) plant produces gametes (p. 175) by mitosis (p. 37) which fuse to form a zygote (p. 166) that develops into the sporophyte (↑).

archegonium (n) the female sex organ of Hepaticae (liverworts) (p. 52), Musci (mosses) (p. 52), Filicales (ferns) (p. 56) and most gymnosperms (p. 57). It is a multicellular (p. 9) structure which is shaped like a flask with a narrow neck and a swollen base which contains the female gamete (p. 175).

oosphere (n) the large, unprotected, non-motile female gamete (p. 175) found in an archegonium (↑).

ovum[p] (*n*) the haploid (p.36) female gamete (p.175).

antheridium (*n*) the male sex organ of algae (p. 44), liverworts (p. 52), mosses (p. 52), ferns (p. 56) and fungi (p. 46). It may be unicellular (p. 9) or multicellular (p. 9) and produces small, motile gametes (p. 175) – the antherozoids (↓). **antheridia** (*pl.*).

antherozoid (*n*) the male gamete (p. 175) produced within the antheridium (↑).

spermatozoid (*n*) = antherozoid (↑).

sporogonium (*n*) the sporophyte (p. 177) generation of mosses (p. 52) and liverworts (p. 52) which produces the seed and parasitizes (p. 110) the gametophyte (p. 177) generation.

sporangium (*n*) (1) the organ in which, in the sporophyte (p. 177) generation, the haploid (p. 36) spores (↓) are formed after meiotic (p. 38) division of the spore mother cells (↓): (2) in fungi (p. 46) a swelling occurring at the ends of certain hyphae (p. 46) in which protoplasm (p. 10) breaks up to form spores during asexual reproduction (p. 173). **sporangia** (*pl.*).

spore (*n*) a tiny, asexual, unicellular (p. 9) or multicellular (p. 9) reproductive (p. 173) body which is produced in vast numbers by fungi (p. 46) or the sporangia (↑) of plants.

spore mother cell a diploid (p. 36) cell that gives rise to four haploid (p. 36) cells by meiosis (p. 38). Also known as **sporocyte**.

microsporangium (*n*) a sporangium (↑) present in heterosporous (p. 54) plants, which produces and disperses the microspores (↓).

microspore (*n*) the smaller of the two different kinds of spore (↑) produced by ferns (p. 56) and spermatophytes (p. 57) and which gives rise to the male gametophyte (p. 177) generation.

microsporophyll (*n*) a modified leaf that bears the microsporangium (↑).

megasporangium (*n*) a sporangium (↑) present in heterosporous (p. 54) plants, which produces and disperses the megaspores (↓).

megaspore (*n*) the larger of the two different kinds of spore (↑) produced by ferns (p. 56) and spermatophytes (p. 57) and which gives rise to the female gametophyte (p. 177) generation.

heterosporous plants
the production of megaspores

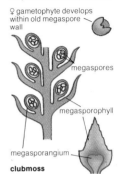

♀ gametophyte develops within old megaspore wall

megaspores

megasporophyll

megasporangium

clubmoss

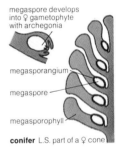

megaspore develops into ♀ gametophyte with archegonia

megasporangium

megaspore

megasporophyll

conifer L.S. part of a ♀ cone

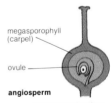

megasporophyll (carpel)

ovule

angiosperm

megaspore develops into ♀ gametophyte (embryo sac)

megaspore

young ovule

flower

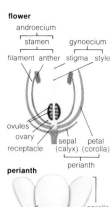

andraecium
stamen
gynoecium
filament anther stigma style
ovules
ovary
receptacle
sepal petal
(calyx) (corolla)
perianth

perianth

corolla
petals
sepals
generalized
flower
calyx

gynoecium

stigma
style
locule
ovary,
containing
ovules

types of ovary

megasporophyll (n) a modified leaf that bears the megasporangium (↑). They can be grouped in a strobilus (p. 55).

flower (n) the structure concerned with sexual reproduction (p. 173) in angiosperms (p. 57). It is a modified vegetative shoot.

corolla (n) the part of the flower which is made up of all the petals (↓). It varies considerably in size, shape, form and colour and often attracts insects (p. 69) to visit the flower, pollinating (p. 183) the plant in the process.

petal (n) one of the often brightly coloured and scented individual elements which make up the corolla (↑). They are thought to be modified leaves. Those flowers which are pollinated (p. 183) by the wind have petals which are greatly reduced in size or absent.

calyx (n) the outermost part of a flower which comprises a number of sepals (↓) that protect the flower while it is developing in the bud stage.

sepal (n) the usually green, often hairy, leaf-like structures which make up the calyx (↑).

perianth (n) the part of the flower, comprising the corolla (↑) and calyx (↑), which surrounds the stamens (p. 181) and carpels (↓).

gynoecium (n) the female reproductive (p. 173) structure of a flower which is made up of the carpels (↓).

carpel (n) one of the single or more individual female reproductive (p. 173) structures of a plant which make up the gynoecium (↑). Each carpel is made up of an ovary (p. 180), a style (p. 181) and a stigma (p. 181). If there is more than one carpel, they may be fused together or separate.

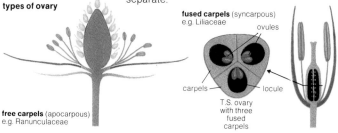

free carpels (apocarpous)
e.g. Ranunculaceae

fused carpels (syncarpous)
e.g. Liliaceae
ovules
carpels
locule
T.S. ovary
with three
fused
carpels

ovary[p] (*n*) the part of the carpel (p. 179) which contains the ovules (↓).

ovule[p] (*n*) the structure containing the female gametes (p. 175) and which, after fertilization (p. 175), becomes the seed.

funicle (*n*) the stalk which attaches the base of the ovule (↑) to the wall of the carpel (p. 179).

placenta[p] (*n*) the part of the wall of the ovary (↑) to which the ovules (↑) are attached.

apocarpous (*adj*) of a gynoecium (p. 179) in which the carpels (p. 179) are not fused.

syncarpous (*adj*) of a gynoecium (p. 179) in which the carpels (p. 179) are fused.

placentation (*n*) the position and arrangement of the placentas (↑) in a syncarpous (↑) gynoecium (p. 179).

parietal (*adj*) of placentation (↑) in which the carpels (p. 179) are fused only by their margins with the placentas (↑) becoming ridges on the inner side of the wall of the ovary (↑).

axile (*adj*) of placentation (↑) in which the carpels (p. 179) fold inwards at their margins, fuse, and become a central placenta (↑).

free central of placentation (↑) in which the placenta (↑) grows upwards from the base of the ovary (↑).

nucellus (*n*) the central tissue (p. 83) of the ovule (↑) enclosing the megaspore (p. 178) or ovum (p. 178).

micropyle (*n*) a canal through the integument (↓) near the apex of the ovule (↑) which, in the seed, becomes a pore (p. 120) through which water may enter to enable germination (p. 168).

integument (*n*) the outermost layer of the ovule (↑) which forms the seed coat.

chalaza (*n*) the base of the ovule (↑) to which the funicle (↑) is attached. It is situated at the point where the nucellus (↑) and integuments (↑) merge.

embryo sac a large, oval-shaped cell, surrounded by a thin cell wall (p. 8), in the nucellus (↑) in which fertilization (p. 175) of the ovum (p. 178) takes place and the embryo (p. 166) develops.

polar nuclei a pair of haploid (p. 36) nuclei (p. 13) found towards the centre of the embryo sac (↑).

egg cell the female gamete (p. 175).

ovule structure

embryo sac
(♀ gametophyte)

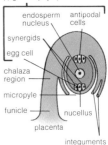

endosperm nucleus
antipodal cells
synergids
egg cell
chalaza region
micropyle
funicle
nucellus
placenta
integuments

placentation types
ovaries cut through to show internal structure

axile

locules

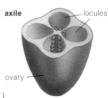

ovary

parietal

locule

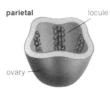

ovary

free-central

locule

ovary

male floral parts
pollen sac   pollen grains

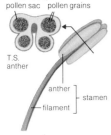

T.S.
anther

anther ⎱
          ⎰ stamen
filament ⎰

**pollen grain of
angiosperm**

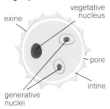

exine

vegetative
nucleus

pore

intine

generative
nuclei

actinomorphic flower
(radial symmetry)

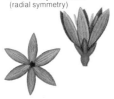

zygomorphic flower
(bilateral symmetry)

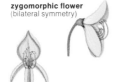

synergid (*n*) one of the two haploid (p. 36) cells which occur at the micropyle (↑) end of the embryo sac (↑) near the egg cell (↑).

antipodal cell one of the three haploid (p. 36) cells that move to the end of the embryo sac (↑) nearest the chalaza (↑). They do not take part in fertilization (p. 175).

style (*n*) the part of the carpel (p. 179) which joins the ovary (↑) and the stigma (↓).

stigma (*n*) the receptive tip of the carpel (p. 179) to which pollen (↓) becomes attached during pollination (p. 183).

androecium (*n*) the male reproductive (p. 173) structure of a flower which is made up of the stamens (↓).

stamen (*n*) one of the male reproductive (p. 173) structures which make up the androecium (↑). A stamen consists of an anther (↓) and a filament (↓).

anther (*n*) the tip of the stamen (↑) which produces the pollen (↓) contained in pollen sacs (↓).

filament (*n*) (1) the stalk of the stamen (↑) to which the anther (↑) is attached; (2) in plants, a chain of cells, e.g. some green algae (p. 44) are filamentous; (3) in animals, any fine threadlike structure.

pollen sac the chamber in which pollen is formed.

pollen (*n*) grain-like microspores (p.178) produced in the pollen sac (↑) in huge numbers by meiotic (p. 38) division of their spore mother cells (p. 178). They contain the male gametes (p. 175).

tapetal cell one of the layer of cells which surrounds the spore mother cells (p. 178) and which provide nutrients for the spore mother cells and the developing spores (p. 178).

generative nucleus one of the two nuclei (p. 13) found in each grain of pollen (↑), both of which are transferred to the ovule (↑) via growth of the pollen tube (p. 184).

receptacle (*n*) the part of the stem of a flower which is often expanded and which bears the organs of the flower.

zygomorphic (*adj*) of a flower, such as a snapdragon, which is bilaterally symmetrical (p. 62).

actinomorphic (*adj*) of a flower, such as a buttercup, which is radially symmetrical (p. 60).

inflorescence
e.g. capitate inflorescence of a composite

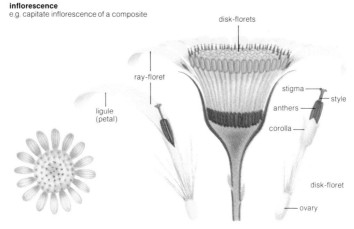

disk-florets

ray-floret

ligule
(petal)

stigma

style

anthers

corolla

disk-floret

ovary

unisexual (*adj*) of a flower, which has the stamens
(p. 181) and carpels (p. 179) on separate
flowers. Unisexual flowers may be either
monoecious (p. 175) or dioecious (p. 175).

nectary (*n*) a glandular (p. 87) swelling found on
the receptacle (p. 181) or other parts of some
flowers, which produces nectar (↓).

nectar (*n*) a sweet, sugary solution (p. 118)
produced by the nectaries (↑). Many insects
(p. 69) visit flowers which produce nectar to feed
on it and, in so doing, pollinate (↓) the flower.

inflorescence (*n*) a group of flowers sharing the
same stem.

inflorescence types

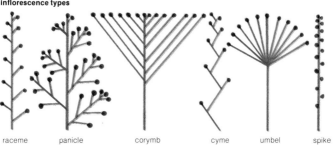

raceme   panicle   corymb   cyme   umbel   spike

spikelet

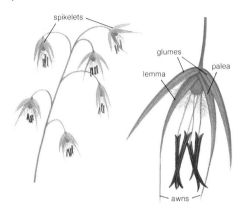

spikelets

glumes

lemma

palea

awns

spathe

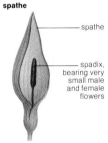

spathe

spadix, bearing very small male and female flowers

cross-pollination

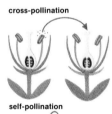

self-pollination

spikelet (*n*) the inflorescence (↑) of a grass.

spathe (*n*) a leaf-like structure or bract which encloses the spadix (↓) of certain monocotyledonous (p. 58) flowers.

spadix (*n*) the inflorescence (↑) of certain monocotyledonous (p. 58) flowers, bearing unisexual (↑) or hermaphrodite (p. 175) flowers.

floral formula a 'shorthand' way of describing the structure of a flower. It is given by a combination of capital letters and numbers as follows: K = calyx (p. 179); C = corolla (p. 179); A = androecium (p. 181); and G = gynoecium (p. 179). Thus, a flower with the floral formula K6 C6 G1 A5 would have six sepals, six petals, one carpel and five stamens.

pollination (*n*) the process in which pollen (p. 181) is transferred from the anther (p. 181) to the stigma (p. 181). **pollinate** (*v*).

self-pollination (*n*) pollination (↑) within the same flower or flowers from the same plant.

cross-pollination (*n*) pollination (↑) between flowers of different plants.

wind pollination pollination (↑) in which the pollen (p. 181) is carried from one flower to the next by the wind.

anemophily = wind pollination (↑).

insect pollination pollination (p. 183) in which the pollen (p. 181) is transferred from one flower to the next on the bodies of insects (p. 69) which are attracted to the flowers by the brightly coloured petals (p. 179), the scent, and the promise of nectar (p. 182).

entomophily = insect pollination (↑).

pollen tube a tubular outgrowth which forms when the pollen (p. 181) grain germinates (p. 168) and which is the means through which male gametes (p. 175) are carried to the egg.

double fertilization in flowering plants, the union of one generative nucleus (p. 181) with an ovum (p. 178) to form a zygote (p. 166) and the other with the two polar nuclei (p. 180) to form the primary endosperm (p. 168) nucleus (p. 13) which is triploid (p. 207). Subsequent division of the primary endosperm nucleus produces the endosperm.

seed (*n*) the structure which develops after the fertilization (p. 175) of the ovule (p. 180) and which is made up of the testa (p. 168) surrounding the embryo (p. 166). In suitable conditions, each seed may germinate (p. 168) and form a fully independent plant. Seeds in flowering plants may be contained in a fruit.

fruit (*n*) the ripened ovary (p. 180) wall of a flower which contains the seeds. Depending upon the method by which the seeds of the plant are distributed, the fruit may be fleshy (distributed by animals) or dry (distributed by wind or water).

pericarp (*n*) the outer wall of the ovary (p. 180) which develops into the fruit.

endocarp (*n*) the inner layer of the pericarp (↑) which develops into the stony covering of the seed of a drupe (↓), such as a cherry.

mesocarp (*n*) the middle layer of the pericarp (↑) which can form the fleshy part of a drupe (↓), such as a cherry, or the hard shell of a nut, like an almond.

exocarp (*n*) the tough outer 'skin' of a fruit.

epicarp (*n*) = exocarp (↑).

aleurone layer the outermost, protein-rich layer of the endosperm (p. 168) of the seeds of grasses.

fertilization in angiosperms

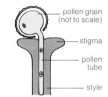

1 pollen grain lands on stigma, pollen tube grows through tissues of style carrying the male gametes

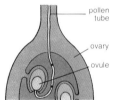

2 pollen tube grows through ovary wall and into micropyle of ovule

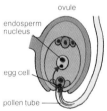

3 one male gamete fertilizes egg cell, the other fertilizes the endosperm nucleus forming endosperm mother cell

fruits and fruit structure

berry e.g. tomato

- exocarp
- seeds
- mesocarp
- endocarp

drupe e.g. apricot

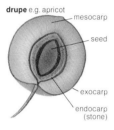

- mesocarp
- seed
- exocarp
- endocarp (stone)

legume e.g. pea

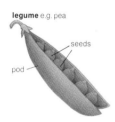

- seeds
- pod

achene e.g. strawberry

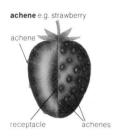

- achene
- receptacle
- achenes

scutellum (n) the part of the embryo (p. 166) of a grass seed, which is situated next to the endosperm (p. 168).

coleoptile (n) the protective sheath with a hard pointed tip which protects the plumule (p. 168) in a germinating (p. 168) grass seedling.

dehiscence (n) the process in which the mature fruit wall opens, sometimes violently, to release the seeds. **dehisce** (v).

berry (n) a fruit, such as a blackberry, which, unlike a drupe (↓) does not have a stony endocarp (↑) so that the seeds are surrounded by a fleshy mesocarp (↑) and endocarp.

drupe (n) a fruit, such as a plum, formed from a single carpel (p. 179) which has a stony endocarp (↑) that surrounds the seed.

follicle[D] (n) a dry fruit, such as a delphinium, which has formed from a single carpel (p. 179) and in which, during dehiscence (↑), the fruit or pod splits along one line to release the seed.

legume (n) a dry fruit, such as a pea, which has formed from a single carpel (p. 179) and in which, during dehiscence (↑), the fruit or pod splits along two sides to release the seed.

siliqua (n) a dry, elongated fruit or special type of capsule (p. 53), such as that found in the cabbage family, formed from two carpels (p. 179) which are fused together but separated by a false septum or wall. During dehiscence (↑), the siliqua splits as the carpel walls separate leaving the seeds attached to the septum.

silicula (n) a type of siliqua (↑), found in plants such as the shepherd's purse, which is short and broad in shape.

achene (n) a dry fruit, such as that of the buttercup, which is formed from a single carpel (p. 179), contains only one seed, has a leathery pericarp (↑), and has no particular method of dehiscence (↑).

cypsela (n) a dry fruit, such as that of the dandelion, which is formed from two carpels (p. 179) of an inferior ovary (p. 180) which retains a plumed calyx (p. 179) to aid in wind dispersal (p. 186).

caryopsis (*n*) a dry fruit, such as that of grasses, which is similar to an achene (p. 185) except that the pericarp (p. 184) is united with the testa (p. 168).

nut (*n*) a dry fruit, such as that of the hazel, which is similar to an achene (p. 185), except that the pericarp (p. 184) is stony.

samara (*n*) a dry fruit, such as that of the elm, which is similar to an achene (p. 185), except that part of the pericarp (p. 184) forms wings that aid in wind dispersal (↓).

false fruit a fruit that includes other parts of the flower, such as the inflorescence (p. 182), as well as the ovary (p. 180).

pome (*n*) a fleshy false fruit (↑), such as that of the apple. The main flesh is made of the swollen receptacle (p. 181).

fruit dispersal the various methods by which a flower distributes its seeds and which may include the fruit or just the seeds alone.

mechanical dispersal fruit dispersal (↑) in which the fruit itself is responsible for distributing the seed by opening explosively when the seed is mature and scattering it widely.

wind dispersal fruit dispersal (↑) in which the seed is carried on the wind either because it is small and lightweight or by bearing wing-like structures which give them extra lift. In some cases, such as that of the poppy, the capsule (p. 53) itself sways in the wind to distribute the seed, like a censer.

animal dispersal fruit dispersal (↑) in which the seed is distributed by being transported by animals, including humans. The seed or fruit may have hooks or spines which stick to the animals' coats or their fruits may be palatable while the seeds may be indigestible so that the fruits are eaten and the seeds pass through undamaged. Indeed, some seeds can only germinate (p. 168) when they have passed through the digestive (p. 98) system of certain animals.

water dispersal fruit dispersal (↑) in which the fruit or seed is specially adapted to be carried in running water.

caryopsis e.g. wheat

nut e.g. hazelnut

samara e.g. sycamore
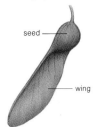
seed

wing

pome e.g. apple
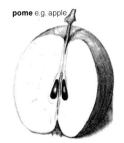

gonad (*n*) the male or female organ of reproduction (p. 173) in sexual animals that produce gametes (p. 175). In some cases, the gonads also produce hormones (p. 130).

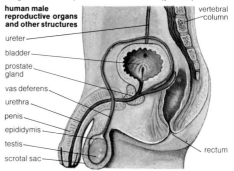

human male reproductive organs and other structures

- vertebral column
- ureter
- bladder
- prostate gland
- vas deferens
- urethra
- penis
- epididymis
- testis
- scrotal sac
- rectum

spermatogenesis

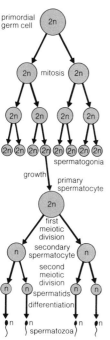

primordial germ cell — 2n

mitosis — 2n / 2n

2n 2n 2n 2n

2n 2n 2n 2n 2n 2n 2n 2n
spermatogonia

growth — primary spermatocyte

2n

first meiotic division

secondary spermatocyte — n / n

second meiotic division

spermatids — n n n n

differentiation

spermatozoa — n n n n

testis (*n*) the male reproductive (p. 173) organ which produces spermatozoa (p. 188) by spermatogenesis (↓). In vertebrates (p. 74) there are two testes which usually lie in a sac of skin, the scrotum or scrotal sac (p. 188), outside the main body cavity and behind the penis (p. 189). In vertebrates, the testes also produce androgens (p. 195).

testicle (*n*) = testis (↑).

seminiferous tubule one of several hundred tiny coiled tubes which compose the testis (↑) and in which all stages of spermatogenesis (↓) take place.

Sertoli cell one of a number of large, specialized cells made of germinal epithelium (p. 87) and found in the testis (↑) which are thought to nourish the spermatids (p. 188) to which they are attached.

spermatogenesis (*n*) the process by which spermatozoa (p. 188) are produced in the testes (↑). A germ cell (p. 36) divides a number of times during a multiplication phase to produce spermatogonia (p. 188) each of which then grows into a primary spermatocyte (p. 188). In turn, the spermatocyte undergoes two phases of meiotic (p. 38) division to produce spermatids (p. 188) which then differentiate into the spermatozoa.

spermatogonium (*n*) one of the large numbers of cells found in the testes (p. 187) and which grows into a primary spermatocyte (↓) during spermatogenesis (p. 187). **spermatogonia** (*pl.*).

spermatocyte (*n*) one of the large numbers of reproductive cells found in the seminiferous tubules (p. 187) and which is produced by the growth of a spermatogonium (↑).

spermatid (*n*) one of the large numbers of reproductive (p. 173) cells found in the seminiferous tubules (p. 187) during spermatogenesis (p. 187). It is produced as a result of two phases of meiosis (p. 38) of a spermatocyte (↑). Each spermatid, nourished by the Sertoli cells (p. 187), differentiates and matures into a spermatozoon (↓).

spermatozoon (*n*) the small, differentiated, highly motile mature male gamete (p. 175) or reproductive (p. 173) cell. Spermatozoa (*pl.*) are produced continuously in large numbers in the seminiferous tubules (p. 187). Locomotion (p. 143) takes place by movements of a flagellum (p. 12).

vas efferens one of the small channels through which spermatozoa (↑) are transported from the seminiferous tubules (p. 187) to the epididymis (↓).

epididymis (*n*) a muscular (p. 143) coiled tubule between the vas efferens (↑) and the vas deferens (↓) which functions as a temporary storage vessel for spermatozoa (↑) until they are released during mating.

vas deferens one of the pair of muscular (p. 143) tubules, with mucous (p. 99) glands (p. 87), which leads from the epididymis (↑) and through which spermatozoa (↑) are released into the urethra (↓) during mating.

urethra (*n*) a duct which leads from the bladder (p. 135) to the exterior and through which urine (p. 135) is excreted (p. 134). In males it also connects with the vas deferens (↑).

scrotal sac the external sac of skin which is divided into two, each carrying one testis (p. 187). Thus, the testes are maintained at a lower temperature than the rest of the body to ensure the best conditions for the development of spermatozoa (↑).

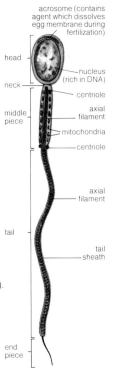

spermatozoon
e.g. human sperm

acrosome (contains agent which dissolves egg membrane during fertilization)

head

nucleus (rich in DNA)

neck

centriole

middle piece

axial filament

mitochondria

centriole

axial filament

tail

tail sheath

end piece

prostate gland a gland (p. 87) surrounding the urethra (↑) which, under the control of androgens (p. 195), secretes (p. 106) alkaline substances that reduce the urine's (p. 135) acidity and aid in the motility of spermatozoa (↑).

seminal vesicle one of the two organs connected to the vas deferens (↑) in most male mammals (p. 80). It is under hormonal (p. 130) control and secretes (p. 106) fluid which makes up the bulk of the semen (p. 191) improving the motility of the spermatozoa (↑).

Cowper's gland one of two glands (p. 87) connected to the vas deferens (↑) which secretes (p. 106) fluid for the semen (p. 191).

penis (*n*) the organ through which the urethra (↑) connects with the exterior and which functions, during mating, to transport spermatozoa (↑) to the female reproductive (p. 173) organs. It contains spongy tissue (p. 83) which fills with blood (p. 90) during mating to become more rigid or erect.

ovary[a] (*n*) one of the pair of female reproductive (p. 173) organs in which ova (p. 190) are produced during oogenesis (↓). Female hormones (p. 130) are also produced in the ovaries.

oogenesis (*n*) the process by which ova (p. 190) are produced in the ovaries (↑). A germ cell (p. 36) divides by mitosis (p. 37) to form a number of oogonia (↓) each of which grows to give rise to a primary oocyte (↓). By two phases of meiotic (p. 38) division – with the second phase usually following fertilization (p. 175) – an ovum is produced together with additional polar bodies (↓).

oogonium (*n*) a specialized cell found within the ovary (↑) which is produced by mitotic (p. 37) division of the germ cell (p. 36) and which grows to give rise to a primary oocyte (↓) during oogenesis (↑). **oogonia** (*pl.*).

oocyte (*n*) a reproductive (p. 173) cell found within the ovary (↑) during oogenesis (↑). It results from the growth of an oogonium (↑).

polar body a tiny cell produced during oogenesis after the second meiotic (p. 38) division when the ovum (p. 190) is formed. The polar body contains a nucleus (p. 13) but virtually no cytoplasm (p. 10).

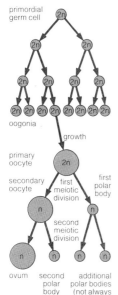

oogenesis

primordial germ cell

2n

2n 2n

2n 2n 2n 2n

2n 2n 2n 2n 2n 2n 2n 2n

oogonia

growth

primary oocyte — 2n

secondary oocyte — first meiotic division — first polar body

n — n

second meiotic division

n — n n n

ovum — second polar body — additional polar bodies (not always formed)

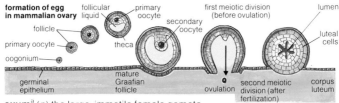

formation of egg in mammalian ovary follicular liquid, primary oocyte, secondary oocyte, first meiotic division (before ovulation), lumen, luteal cells, theca, primary oocyte, follicle, oogonium, germinal epithelium, mature Graafian follicle, ovulation, second meiotic division (after fertilization), corpus luteum

ovum[a] (*n*) the large, immotile female gamete (p. 175) produced in the ovary (p. 189) during oogenesis (p. 189). If it is fertilized (p. 175) by a spermatozoon (p. 188) it develops into a new individual. Fertilization may take place at the oocyte (p. 189) stage following the first meiotic (p. 38) division. **ova** (*pl.*).

Graafian follicle a fluid-filled, spherical mass of cells with a cavity that is found in the ovary (p. 189) and contains an oocyte (p. 189) attached to its wall. It is the site of development of the ovum (↑) and grows from one of the large number of follicles within the ovary.

corpus luteum a gland (p. 87) which forms temporarily in the Graafian follicle (↑) after rupture during ovulation (p. 194). It secretes (p. 106) the hormone (p. 130) progesterone (p. 195) which, if the ovum (↑) is fertilized (p. 175), continues to be released to prepare the female reproductive (p. 173) tract for pregnancy (p. 195). If fertilization does not take place, the corpus luteum degenerates. **corpora lutea** (*pl.*).

oviduct (*n*) a muscular (p. 143) tube lined with cilia (p. 12) by which ova (↑) are transported from the ovaries (p. 189) to the exterior.

uterus (*n*) a thick-walled organ in which the embryo (p. 166) develops. It is muscular (p. 143) and the smooth muscle increases in amount during pregnancy (p. 195) so that it is able to expel the young at birth. The size of the uterus as well as the thickness of its wall, which provides a point of attachment and nourishment for the developing embryo, varies cyclically and with sexual activity or inactivity under the influence of reproductive (p.173) hormones (p. 130). Also known as **womb**.

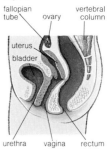

human female reproductive organs

fallopian tube, ovary, vertebral column, uterus, bladder, urethra, vagina, rectum

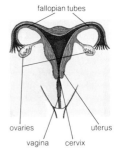

fallopian tubes, ovaries, uterus, vagina, cervix

cervix (*n*) a ring of muscle (p. 143) between the uterus (↑) and the vagina (↓) which also contains mucous glands (p. 87).

vagina (*n*) the muscular (p. 143) duct which connects the uterus (↑) to the exterior and which receives the penis (p. 189) during mating.

copulation (*n*) the sexual union of male and female animals during mating in which, in mammals (p. 80), the penis (p. 189) is received by the vagina (↑) and ejaculation (↓) takes place. Also known as **coitus**.

semen (*n*) a fluid containing spermatozoa (p. 188) produced by the testes (p. 187) and other liquids produced by the prostate gland (p. 189). During copulation (↑) semen is passed from male to female.

ejaculation (*n*) the rhythmic and forcible discharge of semen (↑) from the penis (p. 189).

orgasm (*n*) the climax of sexual excitement which takes place during mating and involves a complex series of reactions of the reproductive (p. 173) organs and other parts of the body including the skin.

implantation (*n*) following fertilization (p. 175), the process in which the developing zygote (p. 166) embeds itself in the wall of the uterus (↑).

foetus (*n*) an embryo (p. 166) with an umbilical cord (p. 192) which is sufficiently developed to show the main features that the mammal (p. 80) will possess after birth.

foetal membrane any one of those membranes (p. 14) or structures which are developed by the embryo (p. 166) for nourishment and protection but which do not form part of the embryo itself.

amnion (*n*) the fluid-filled sac in which the embryo (p. 166) develops in mammals (p. 80). The amnion offers the embryo protection from any pressure exerted on it by the organs of the mother and a liquid environment (p. 218) in which to develop (important for land animals). The sac wall consists of two layers of epithelium (p. 87) and sometimes only the inner layer is referred to as the amnion. **amniotic** (*adj*), **amniote** (*adj*).

amniotic cavity the amnion (↑) or the fluid-filled cavity within the amnion which contains the developing embryo (p. 166).

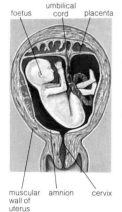

foetus in uterus

foetus   umbilical cord   placenta

muscular wall of uterus   amnion   cervix

allantois (*n*) a sac-like extension of the gut (p. 98) which is present in the embryos (p. 166) of reptiles (p. 78), birds and mammals (p. 80) and which grows out beyond the embryo itself. The connective tissue (p. 88) which covers it is liberally supplied with blood vessels (p. 127) and functions for gas exchange (p. 112) of the embryo as well as for storing the products of excretion (p. 134).

chorion (*n*) the outermost membrane (p. 14), the outer epithelium (p. 87) of the amnion (p. 191) wall which surrounds the embryo (p. 166) of mammals (p. 80) and which unites with the allantois (↑) to develop into the placenta (↓).

placenta[a] (*n*) a disc-shaped organ which develops within the uterus (p. 190) during pregnancy (p. 195) and which is in close association with the embryo (p. 166) and with tissues (p. 83) of the mother. The placenta serves for attachment and nourishment over its large surface area.

umbilical cord a cord which connects the placenta (↑) to the navel of the foetus (p. 191) allowing interchange of materials via two arteries (p. 127) and a vein (p. 127).

viviparity (*n*) the condition in which embryos (p. 166) develop within a uterus (p. 190), are attached to a placenta (↑), and are born alive. **viviparous** (*adj*).

gestation period the time which elapses between fertilization (p. 175) of the ovum (p. 190) and the birth of the young in viviparous (↑) animals. It varies from species (p. 40) to species.

parturition (*n*) the process of giving birth to live young in viviparous (↑) animals by rhythmic contractions stimulated by the secretion (p. 106) of certain hormones (p. 130).

lactation (*n*) the production of milk in the mammary glands (p. 87) to nourish the young in mammals (p. 80).

puberty (*n*) the sexual maturity of a mammal (p. 80).

menopause (*n*) the period in females during which the menstrual cycle (p. 194) becomes irregular with increasing age of the individual before ceasing totally.

embryonic membranes of a mammal

embryo
chorion
amnion
embryonic gut
yolk sac
allanto-chorionic placenta villi
yolk sac placenta villi
allantois

embryonic membranes in a reptilian egg

amniotic fluid
chorion
shell
amnion
yolk
yolk sac
allantois
embryo

sexual cycle the sequence of events which occurs in the females of animals that reproduce (p. 173) sexually and which, in humans, takes place on a monthly pattern with menstruation (p. 194) alternating with ovulation (p. 194).

oestrus cycle the rhythmic sexual cycle (↑) which occurs in mature females of most mammals (p. 80) assuming that the female does not become pregnant (p. 195). There are four main events in the oestrus cycle of which the most important is oestrus (p. 194) itself. In the *follicular phase*, there is growth of the Graafian follicles (p. 190), a thickening of the lining of the uterus (p. 190) and an increase in the production of oestrogen (p. 194). This is followed by *oestrus*. Then comes the *luteal phase* during which a corpus luteum (p. 190) grows from the Graafian follicle which secretes (p. 106) progesterone (p. 195) with a reduction in the secretion of oestrogen. If fertilization (p. 175) and pregnancy occur then the cycle is interrupted and the fourth phase does not follow. If fertilization does not occur, then the corpus luteum diminishes, hormone (p. 130) levels fall, and a new Graafian follicle begins to grow.

relationships between hormones secreted by pituitary, the oestrus cycle and pregnancy in a human

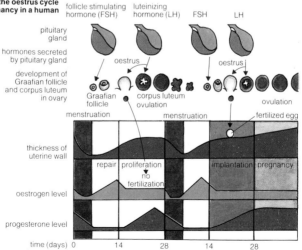

follicle stimulating hormone (FSH) luteinizing hormone (LH) FSH LH

pituitary gland

hormones secreted by pituitary gland

oestrus oestrus

development of Graafian follicle and corpus luteum in ovary

Graafian follicle corpus luteum ovulation ovulation

menstruation menstruation fertilized egg

thickness of uterine wall

repair proliferation no fertilization implantation pregnancy

oestrogen level

progesterone level

time (days) 0 14 28 14 28

menstrual cycle in humans and some other primates a modified version of the oestrus cycle (p. 193) in which the oestrus (↓) is not obvious so that the female is continuously attractive and receptive to males. There is a regular discharge of blood (p. 90) and the lining of the uterus (p. 190) (menstruation) which occurs, following ovulation (↓) when fertilization (p. 175) does not occur.

ovulation (*n*) the release from the Graafian follicle (p. 190) of an immature ovum (p. 190) or oocyte (p. 189). It takes place, under the influence of a hormone (p. 130) released by the pituitary gland (p. 157) at regular intervals (approximately every 28 days in humans) and in the presence of oestrogen (↓). **ovulate** (*v*).

oestrus (*n*) a short period during the sexual cycle (p. 193) of animals in which the female ovulates (↑) and is also sexually attractive to males so that copulation (p. 191) takes place.

follicle-stimulating hormone FSH. A hormone (p. 130) produced by the pituitary gland (p. 157), following the completion of the oestrus cycle (p. 193) or pregnancy (↓) which stimulates the growth of the Graafian follicles (p. 190) and the ova (p. 190) in females and spermatogenesis (p. 187) in males.

oestrogen (*n*) a female sex hormone (p. 130), produced in the Graafian follicle (p. 190), which stimulates the production of a suitable environment (p. 218) for fertilization (p. 175) and then growth of the embryo (p. 166) by repairing the walls of the uterus (p. 190) following menstruation (↑). During the first part of the oestrus cycle (p. 191), it builds up until it stimulates the production of luteinizing hormone (↓) by the pituitary gland (p. 157). It is also involved in the development of other female organs associated with the sexual cycle.

luteinizing hormone LH. A hormone (p. 130) secreted (p. 106) by the pituitary gland (p. 157) under the influence of oestrogen (↑). It stimulates ovulation (↑) and development of a Graafian follicle (p. 190) into a corpus luteum (p. 190) which produces progesterone (p. 195).

interaction of hormones in the female sexual cycle

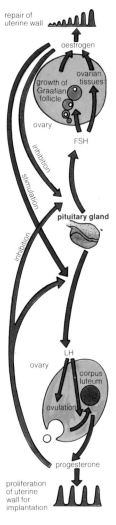

progesterone (*n*) a hormone (p. 130) secreted (p. 106) by the corpus luteum (p. 190) which stops further Graafian follicles (p. 190) from developing by preventing the secretion of follicle-stimulating hormone (↑). It also prepares the uterus (p. 190) for implantation (p. 191) of the ova (p. 190), and assists in the development of the placenta (p.192) and mammary glands (p.87).

pregnancy (*n*) the condition which occurs in a female following successful fertilization (p. 175) and implantation (p. 191). The oestrus cycle (p. 193) is suspended in the luteal phase. The production of hormones (p. 130) is altered so that they are produced by the placenta (p. 192) as well as the pituitary gland (p. 157) to ensure that parturition (p. 192) and lactation (p. 192) take place properly. **pregnant** (*n, adj*).

oxytocin (*n*) a hormone (p. 130) produced by the pituitary gland (p. 157) at the end of pregnancy (↑) which stimulates the contraction of uterine (p. 190) muscles (p. 143) during labour and prepares the mammary glands (p. 87) for the production of milk during lactation (p. 192).

prolactin (*n*) a hormone (p. 130) produced by the pituitary gland (p. 157) which stimulates and controls the production of milk during lactation (p. 192).

breeding season in animals in which the oestrus cycle (p. 193) does not occur continuously throughout the year, the time during which it does take place and which is usually under the influence of climate or other environmental (p. 218) factors.

androgens (*n.pl.*) the male sexual hormones (p. 130), such as testosterone (↓), produced essentially by the testes (p. 187), and which stimulate and control spermatogenesis (p. 187) as well as other male characteristics, such as the growth of facial hair.

testosterone (*n*) an androgen (↑) produced by male vertebrates (p. 74).

interstitial cell-stimulating hormone a luteinizing hormone (↑) which stimulates the secretion (p. 106) of androgens (↑) by the testes (p. 187) in males.

genetics (*n*) the study or science of inheritance concerning the variations between organisms and how these are affected by the interaction of environment (p. 218) and genes (↓).

inherit (*v*) to receive genetic (↓) material from one's parents or ancestors. **inheritance** (*n*).

genotype (*n*) the actual genetic (↓) make-up of an organism which may, for example, define the limits of its growth that are then affected by the environment (p. 218).

phenotype (*n*) the total characteristics and appearance of an organism. Organisms may have the same genotype (↑) while the phenotypes may be different because of the effects of the environment (p. 218).

genome (*n*) the genetic (↓) material.

gene (*n*) the smallest known unit of inheritance that controls a particular characteristic of an organism, such as eye colour. A gene may be considered to be a complex set of chemical compounds sited on a chromosome (p. 13). A gene may replicate to produce accurate copies of itself or mutate (p. 206) to give rise to new forms. **genetic** (*adj*).

Mendelian genetics the system of genetics (↑) developed by the Austrian monk, Gregor Mendel (1822–84), in which he studied inheritance by a series of controlled breeding experiments with the garden pea. He studied simple characteristics, controlled by a single gene (↑), and, using statistics, analyzed the results of cross breeding. In this way he showed that phenotypes (↑) did not result from a blending of genotypes (↑) but that the phenotypes were passed on in different ratios.

first filial (F₁) generation the first generation of offspring resulting from cross breeding pure lines (↓) or parentals (↓) of a single species (p. 40).

second filial (F₂) generation the generation of offspring resulting from the cross breeding between individuals of the first filial generation (↑).

pure line the succession of generations which results from the breeding of a homozygous (↓) organism so that they breed true and produce genetically (↑) identical offspring.

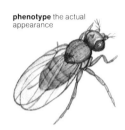

phenotype the actual appearance

genotype the actual genetic makeup as determined by the chromosomes

chromosome

parental (*n*) the succession of generations which leads to filial generations (↑).

monohybrid inheritance the result of cross breeding from pure lines (↑) with one pair of contrasting characteristics to give offspring with one of the characteristics, such as Mendel's cross of tall and dwarf garden peas to give a tall monohybrid.

dominant (*adj*) of (1) a gene (↑) which gives rise to a characteristic that always appears in either a homozygous (↓) or a heterozygous (p. 198) condition e.g. in Mendel's cross of tall and dwarf garden peas, all the F_1 generation (↑) were tall while, in the F_2 generation (↑), tall individuals were in a ratio of 3:1 to dwarf individuals. Thus, the dominant gene was for tallness. (2) a plant species (p. 40) which, in any particular community (p. 217) of plants, is the most common and characteristic species of that community in its numbers and growth. The dominant species has a direct effect on the other plants in the community.

recessive (*adj*) of a gene (↑) which gives rise to a characteristic that can only appear in a homozygous (↓) condition and is suppressed by the dominant (↑) gene in the heterozygous (p. 198) condition. For example, in Mendel's cross of tall and dwarf garden peas, the recessive gene was for dwarfness.

allele (*n*) one of the alternative forms of a gene (↑) e.g. from the pair of genes designated BB giving rise to brown eyes and the pair of genes designated bb giving rise to blue eyes, the genes B and b are said to be alleles of the same gene and B is dominant (↑) while b is recessive (↑).

homozygous (*adj*) of an organism which has the same two alleles (↑) for a particular characteristic, such as eye colour. If a homozygote is crossed with a similar homozygote, it breeds true for that characteristic. If an organism is homozygous for every characteristic and breeds with a genetically (↑) identical organism, the offspring will be identical to the parents. This gradually occurs with constant inbreeding so that, while the organisms may well be adapted to their particular environment (p. 218), if it should change, they would be slow to respond.

alleles

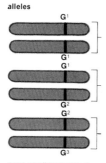

G^1
G^1
G^1
G^1
G^2
G^2
G^3

some possible combinations of 3 alleles on a chromosome pair

heterozygous (*adj*) of an organism which has two different alleles (p. 197) for a particular characteristic, such as eye colour, so that the dominant (p. 197) allele is expressed in the phenotype (p. 196). If a heterozygote breeds with a genetically (p. 196) identical heterozygote, some recessive (p. 197) characteristics will appear in some of the offspring. Heterozygous organisms are more adaptable to changing conditions than homozygous (p. 197) ones.

law of segregation Mendel's first law. One of the two laws formulated by the Gregor Mendel, to explain the way in which inheritance occurred. It states that in two alleles (p. 197) on a gene (p. 196) for a pair of characters, only one can be carried in a single gamete (p. 175).

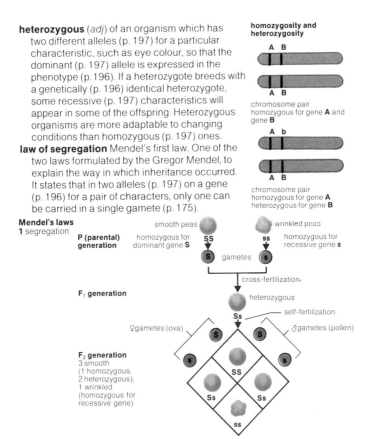

homozygosity and heterozygosity

chromosome pair homozygous for gene **A** and gene **B**

chromosome pair homozygous for gene **A** heterozygous for gene **B**

Mendel's laws
1 segregation

P (parental) generation homozygous for dominant gene **S**

smooth peas

wrinkled peas

homozygous for recessive gene **s**

gametes

cross-fertilization

F₁ generation

heterozygous

self-fertilization

♀gametes (ova)

♂gametes (pollen)

F₂ generation
3 smooth
(1 homozygous,
2 heterozygous),
1 wrinkled
(homozygous for
recessive gene)

test cross a test to show whether or not an organism, which shows a characteristic associated with a dominant (p. 197) gene (p. 196), is heterozygous (↑) or homozygous (p. 197) for that characteristic, by crossing it with a double recessive (↓) for the characteristic. If the organism under test is homozygous, all the offspring will show the characteristic of the dominant gene while, if it is heterozygous, half show the dominant character and half the recessive (p. 197).

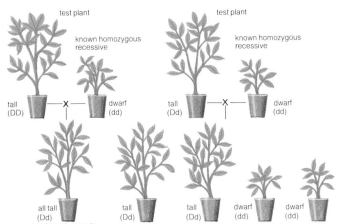

test plant
known homozygous recessive

tall (DD) — X — dwarf (dd)

all tall (Dd)

test plant
known homozygous recessive

tall (Dd) — X — dwarf (dd)

tall (Dd)   tall (Dd)   dwarf (dd)   dwarf (dd)

example of a test cross to see whether a tall individual is heterozygous or homozygous. If homozygous the progeny are all tall; if heterozygous half are tall and half dwarf

double recessive an individual in which the alleles (p. 197) of a particular gene (p. 196) are identical for a recessive (p. 197) characteristic so that the recessive characteristic is expressed in the phenotype (p. 196).

carrier (*n*) an organism which may carry a recessive (p. 197) gene (p. 196) for a characteristic which may be harmful and which is not expressed in the carrier because it is masked by the dominant (p. 197) gene for that characteristic.

dihybrid inheritance the result of cross breeding from pure lines (p. 196) of homozygous (p. 197) organisms with two different alleles (p. 197) for different characteristics, such as Mendel's cross of yellow round and wrinkled green garden peas to give a yellow round dihybrid in which the genes (p. 196) for yellow and round are dominant (p. 197) and suppress the genes for wrinkled and green which are recessive (p. 197).

dihybrid cross the result of dihybrid inheritance (↑). If the offspring of dihybrid inheritance are self crossed, the characteristics are expressed in the ratio 9:3:3:1, in other words, in the Mendel cross, nine plants are yellow and round, three are yellow and wrinkled, three are green and round, and one is green and wrinkled.

law of independent assortment Mendel's
  second law. One of the two laws of inheritance
  formulated by the Gregor Mendel, which states
  that each member of one pair of alleles (p. 197)
  is as likely to be combined with one member of
  another pair of alleles as with any other member
  because they associate randomly (and
  independently).

Mendel's laws
2 independent assortment

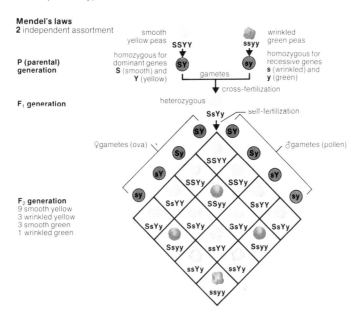

**P (parental)
generation**

F₁ generation

F₂ generation
9 smooth yellow
3 wrinkled yellow
3 smooth green
1 wrinkled green

progeny (*n.pl.*) the offspring which result from
  reproduction (p. 173).
linkage (*n*) the situation in which genes (p. 196)
  on the same chromosome (p. 13) are said to be
  linked so that they are unable to assort
  according to the Law of Independent
  Assortment (↑) and are inherited together.
linkage group a group of linked (↑) genes (p. 196)
  on the same chromosome (p. 13) which are
  inherited together.

sex chromosomes

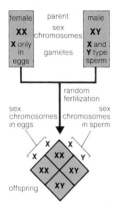

inheritance of colour blindness

X = normal sex chromosome

X^c = sex chromosome with gene for colour blindness

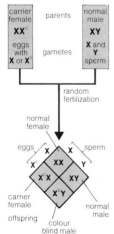

sex chromosomes the chromosomes (p. 13) which control whether or not a given individual of most animals should be male or female. There is a homologous (p. 39) pair of chromosomes in the nucleus (p. 13) of one sex, usually the female, and an unlike or single chromosome in the nucleus of the other, usually the male.

X chromosomes the sex chromosomes (↑) which occur as a like pair XX in the nuclei (p. 13) of the homogametic (↓) sex and usually are responsible for the female sex in most animals. All the gametes (p. 175) of the homogametic sex will contain one X chromosome.

Y chromosomes the sex chromosomes (↑) which occur either as an unlike pair with an X chromosome (↑) or unpaired in the nuclei (p. 13) of the heterogametic (↓) sex and are usually responsible for the male sex in most animals. The gametes (p. 175) of the heterogametic sex are of two kinds, with or without an X chromosome, which are equal in number.

hetersomes (*n.pl.*) homologous chromosomes (p. 39), such as the sex chromosomes (↑) which are not normally identical in appearance.

autosomes (*n.pl.*) homologous chromosomes (p. 39) which are not sex chromosomes (↑) and which are normally identical in appearance.

homogametic sex the sex, usually the female, which contains sex chromosomes (↑) that occur as a like pair of XX chromosomes (p. 13) in the nuclei (p. 13) of an organism.

heterogametic sex the sex, usually the male, which contains sex chromosomes (↑) that occur as an unlike pair of XY chromosomes (p. 13) or unpaired in the nuclei (p. 13) of an organism.

sex-linked (*adj*) of certain characteristics associated with recessive (p. 197) genes (p. 196) which are linked to the sex of the individual because they are attached to the X chromosome (↑).

colour blindness a sex-linked (↑) characteristic in which there is an inability to distinguish between pairs of colours, usually red/green, although the ability to distinguish shade and form is unaffected.

how crossing over and recombination during meiosis shuffles genes and causes variation

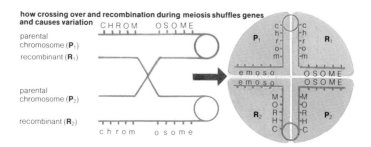

parental chromosome (**P₁**)

recombinant (**R₁**)

parental chromosome (**P₂**)

recombinant (**R₂**)

haemophilia (*n*) a sex-linked (p. 201) characteristic or disease, known only in males, in which the blood (p. 90) is unable to clot (p. 129) properly after wounding.

crossing over the exchange of genetic material (p. 203) during meiosis (p. 39), between male and female parentals (p. 197) in which the chromatids (p. 35) of homologous chromosomes (p. 39) break at the chiasmata (p. 39) and rejoin to allow the assortment of linked (p. 201) genes. Crossing over leads to increased variation (p. 213).

recombinants (*n.pl.*) the gametes (p. 175) which result from crossing over (↑) so that an exchange of genetic material (p. 203) between the parentals (p. 197) gives rise to some characteristics which are not present in either parental. This leads to increased variability in the offspring and greater change of adaptation to changing conditions. **recombine** (*v*).

crossover frequency the number of recombinants (↑) that are likely to occur as a result of the crossing over (↑) between two genes (p. 196) on different parts of the same chromosome (p. 13). It is usually expressed as a percentage of the number of recombinants compared with the total number of offspring produced. The crossover frequency is lower the closer the genes occur on chromosomes.

chromosome map a diagram of the order and distance between the genes (p. 196) on a chromosome (p. 13) worked out by experiments and an analysis of the crossover frequency (↑).

crossing over
during first meiotic division

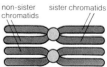

non-sister chromatids   sister chromatids

2 homologous chromosomes

chiasma

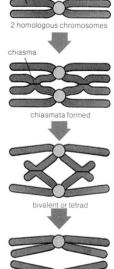

chiasmata formed

bivalent or tetrad

genetic material exchanged, chromosomes separate

locus

homologous chromosomes

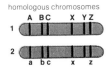

alleles **A**, **B**, **C**, **X**, **Y**, **Z**, occupy
same loci (positions) on
chromosome 1 as alleles **a**, **b**,
c, **x**, **y**, **z**, on chromosome 2

gene locus the precise position of a gene (p. 196)
on the chromosome (p. 13). Alleles (p. 197) of
the same gene occupy the same loci (*pl.*) on
homologous chromosomes (p. 39).

multiple alleles a series of three or more alleles
(p. 197) on the same gene (p. 196) which give
rise to a particular characteristic. Only two of these
alleles in various combinations can occupy
the same gene locus (1) on a pair of homologous
chromosomes (p. 39) at the same time.

lethal alleles alleles (p. 197) which will kill the
individual if they are dominant (p. 197) in a
heterozygous (p. 198) individual or if they are
recessive (p.197) in a homozygous (p.197) one.

partial dominance the situation which occurs
between dominant (p. 197) alleles (p. 1.97) in
which one may be slightly more dominant than
the other. For example, if the allele for red is
dominant in individuals of the same flower
while the allele for white is dominant in other
individuals, breeding them may produce pink
flowers which will be reddish pink if the allele
for red is more dominant than the allele for
white. Also known as **co-dominance**.

epistasis (*n*) the interaction of non-allelic (p. 197)
genes (p. 196) in which one gene suppresses the
characteristics which would normally be expressed
by another gene. It is similar to recessiveness
(p. 197) and dominance (p. 197) between alleles.

genetic material the organic compounds (p. 15)
which carry the genetic (p. 196) information
from one generation to the next and from cell to
cell. Chromosomes (p. 13) are composed of
proteins (p. 21) and DNA (p. 24) which carries
the genetic information.

genetic code the sequence of the four bases (p.22)
adenine (p. 22), guanine (p. 22), cytosine (p.22),
and thymine (p. 22) on a strand of DNA (p. 24)
represents a code that controls the construction of
proteins (p. 21) and enzymes (p. 28) which make
up the cytoplasm (p. 10) of an organism and directs
its functioning. Triplets of these bases code for
the twenty different amino acids (p. 21), and
groups of these triplets, code for whole proteins.
More than one triplet can code for an amino acid.

amino acids and the genetic code
(see previous page)

amino acid general formula	COO^-
	$NH_3{}^+{-}C{-}H$
R = side group	R

codon	amino acid	side group (R)	side group (R)	amino acid	codon
AAA AAG }	lysine	$-CH_2CH_2CH_2CH_2NH_3{}^+$	$-H$	glycine	GGU GGC GGA GGG
AAU AAC }	asparagine	$-CH_2CONH_2$			
ACU ACC ACA ACG	threonine	$-CHOHCH_3$	$-CH_2COO^-$	aspartic acid	GAU GAC
			$-CH_2CH_2COO^-$	glutamic acid	GAA GAG
AGU AGC }	serine	$-CH_2OH$			
AGA AGG }	arginine	$-CH_2CH_2CH_2NHC\overset{NH_2}{\underset{N^+H_2}{}}$	$-CH_3$	alanine	GCU GCC GCA GCG
AUU AUC AUA }	isoleucine	$CH_3CH_2CHCH_3$	CH_3CHCH_3	valine	GUU GUC GUA GUG
AUG }	methionine	$-CH_2CH_2SCH_3$			
CCU CCC CCA CCG	proline	(ring structure)	$-CH_2C$ (benzene ring) CH	phenylalanine	UUU UUC
CAU CAC }	histidine	$-CH_2-C$ (imidazole ring)	$-CH_2-\overset{CH_3}{\underset{CH_3}{CH}}$	leucine	UUA UUG
			$-CH_2C$ (ring) COH	tyrosine	UAU UAC
			NONSENSE		UAA UAG
CAA CAG }	glutamine	$-CH_2CH_2CONH_2$	$-CH_2SH$	cysteine	UGU UGC
CGU CGC CGA CGG	arginine	$-CH_2CH_2CH_2NHC\overset{N^+H_2}{\underset{NH_2}{}}$	$-CH_2C$ (indole ring)	tryptophan	UGG
CUU CUC CUA CUG	leucine	$-CH_2\overset{CH_3}{\underset{CH_3}{CH}}$	NONSENSE		UGA
			$-CH_2OH$	serine	UCU UCC UCA UCG

transcription (*n*) the process in which the genetic
code (p. 203) is, in the first place, copied from
the DNA (p. 24) on to a single strand of RNA
(p. 24) in the nuclei (p. 13) of cells.
translation (*n*) the process in which the
messenger RNA (p. 24) from the transcription
(1) then leaves the nucleus (p. 13) and passes
into the ribosomes (p. 10) in the cytoplasm
(p. 10) to function as a pattern from which amino
acids (p. 21) are built into proteins (p. 21).

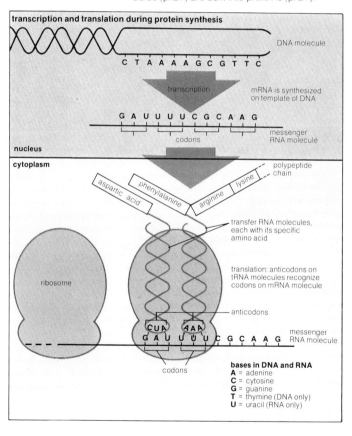

mutation (*n*) a change in the structure of the genetic material (p. 203) of an organism which will be inherited if it occurs in the cells which produce the gametes (p. 175). It can occur as a result of changes in genes (p. 196) or of changes in the structure or number of chromosomes (p. 13). Most mutations are harmful but some allow the organism to adapt to changing circumstances and as a source of increased variation (p. 213) are the very material of evolution (p. 208). Mutations can be stimulated by increases in certain chemicals or ionizing radiation.

mutant (*n*) the result of a mutation (↑) which is usually recessive (p. 197) in the most common types of mutation.

mutagenic agent some stimulus, such as certain chemicals or ionizing radiation, which is likely to cause a mutation (↑).

chromosome mutation a change or mutation (↑) of the number or arrangement of the chromosomes (p. 13).

deletion (*n*) a chromosome (p. 13) mutation (↑) which occurs if a segment of a chromosome breaks away and is lost during nuclear division (p. 35) with a resulting loss of genetic material (p. 203).

inversion (*n*) (1) a chromosome (p. 13) mutation (↑) which occurs if a segment of a chromosome breaks away during nuclear division (p. 35) and rejoins the chromosome the wrong way round to reverse the sequence of genes (p. 196); (2) a gene mutation (↓) in which the order of the bases (p. 22) in a strand of DNA (p. 24) is changed.

translocation[2] (*n*) a chromosome (p. 13) mutation (↑) which occurs if a segment of the chromosome breaks away during nuclear division (p. 35) and rejoins the original chromosome in a different place or joins another chromosome.

duplication (*n*) a chromosome (p. 13) mutation (↑) in which a segment of the chromosome is duplicated either on the same or on another chromosome.

deletion

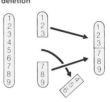

inversion

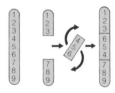

translocation

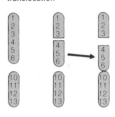

duplication

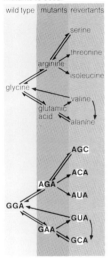

how changes of single nucleotides in a triplet (substitution) can cause mutants to occur

wild type | mutants | revertants

serine
threonine
arginine
isoleucine
glycine
valine
glutamic acid
alanine

AGC
ACA
AGA
AUA
GGA
GUA
GAA
GCA

gene mutation a mutation (↑) in which the sequence of bases (p. 22) is not copied precisely in replicating a strand of DNA (p. 24) resulting in a change in the formation of the proteins (p. 21). Once it has occurred, it is replicated in the formation of further strands of DNA.

substitution (n) a gene mutation (↑) in which one DNA (p. 24) base (p. 22) is replaced by another.

insertion (n) a gene mutation (↑) in which another base (p. 22) is inserted in the existing sequence of bases in the strand DNA (p. 24).

sickle-cell anaemia a disease which is inherited and exhibits partial dominance (p. 203). A sickle cell contains a mutant (↑) gene (p. 196) which crystallizes haemoglobin (p. 126) in the erythrocytes (p. 91) of human blood (p. 90) and distorts them causing the blood vessels (p. 127) to clog. It is found usually among negroid people and is thought to offer some resistance to malaria.

polyploidy (n) the condition in which the cells of an organism contains at least three times the normal haploid (p. 36) number of chromosomes (p. 13). **polyploid** (adj).

triploid (adj) of a polyploid (↑) cell in which there is three times the normal haploid (p. 36) number of chromosomes (p. 13). It results from the fusion of a haploid and a diploid (p. 36) gamete (p. 175).

tetraploid (adj) of a polyploid (↑) cell in which there is four times the normal haploid (p. 36) number of chromosomes (p. 13). It occurs as a result of the fusion of two diploid (p. 36) cells.

aneuploidy (n) the condition in which chromosome (p. 13) mutation (↑) results in the gain or loss of chromosomes from a set.

euploidy (n) the condition in which chromosome (p. 13) mutation (↑) results in the gain of a whole set of chromosomes.

autopolyploid (adj) of the condition of polyploidy (↑) which results from euploidy (↑) in which the cell has multiple sets of its chromosomes (p. 13).

allopolyploid (adj) of the condition of polyploidy (↑) which results from euploidy (↑) in which the cell contains two different sets of chromosomes (p. 13) from the hybridization (p. 216) of two closely related organisms especially plants.

evolution (*n*) the process whereby all organisms descend from the common ancestors which emerged on the Earth. Over successive generations throughout geological time, populations (p. 214) are modified in response to changes in environment (p. 218) by such processes as natural selection (↓) so that new species (p. 40) are formed which are all related, however distantly, by common descent.

Darwinism (*n*) the mechanism first put forward by the British naturalist, Charles Darwin (1809–82), following careful observation of animals and plants all over the world, e.g. Darwin's finches, to explain how organisms changed slowly, over millions of years, to evolve new forms. He suggested that in any given population (p. 214) of an organism there was considerable variation (p. 213) between individuals. Some would exhibit different characteristics which would be better fitted to their circumstances and environment (p. 218) than others. Thus, these individuals would be more likely to survive to maturity and breed so that their offspring would also exhibit these characteristics. Those individuals that were less well suited to their conditions would have less chance of breeding success so that eventually the population would contain more and more of the individuals that were better suited to their environment and the character of the species (p. 40) would change as a whole and result in a new species. Darwin was unable to explain, however, how the variations were produced in the first place. He called this process the theory of natural selection (↓).

natural selection one of the central deductions of Darwinism (↑). If the variations (p. 213), which occur among individuals within a population (p. 214) of animals, gives to those individuals a better chance of surviving, they are more likely to reach sexual maturity and breed so that their offspring will also inherit those advantageous characteristics. Eventually, the inheritance of variations in a particular direction over generations will lead to a new species (p. 40).

Darwin's finches on the Galapagos. Darwin observed that there were many species of finches which he thought had evolved from one species after it had arrived on the islands

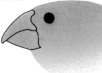

C. crassirostris
vegetarian

C. psittacula
insect feeding

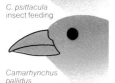

Camarhynchus pallidus
woodpecker finch

G. fortis
ground feeding

Geospiza scandens
cactus feeding

Pinaroloxias inornata
warbler-like

survival of the fittest the idea behind natural selection (↑), which suggests that only the animals which are best fitted to their circumstances will survive in the struggle for existence while those that are less well fitted will tend to perish.

Neodarwinism (*n*) the modern, modified version of Darwinism (↑) which, with the aid of the theories (p. 235) of genetics (p. 196) based on the work of Gregor Mendel (p. 196), seeks to explain the mechanisms for the existence of advantageous variations (p. 213) which may occur naturally in a population (p. 214) of organisms which, because of lack of knowledge at the time, Darwin was unable to account for.

origin of species the theory (p. 235) leading from Darwinism (↑), which was developed by Charles Darwin in a paper published by him in 1859 and entitled *On the Origin of Species by Means of Natural Selection and the Preservation of Favoured Races in the Struggle for Life.* The theory suggests that within a population (p. 214) of one species (p. 40), various factors exist, such as geographical barriers (rivers, oceans, mountains, etc) or specific differences in behaviour (p. 164) which can isolate (p. 214) breeding groups within the population. This tends to maintain the integrity of the genes (p. 196) which carry the variations (p. 213) within the breeding group that are advantageous for the local environment (p. 218). In this way, the genetic differences between one group and another can build up, and over many generations lead to the development of new species, each fitted to its own conditions. This is referred to as speciation (p. 213). *See also* natural selection (↑).

Lamarckism (*n*) a theory (p. 235) which stemmed from the observations of the French biologist, Jean de Lamarck (1744–1829), who noticed that particular organs of an animal could fall into use or disuse if they were needed or not. From this he suggested that these acquired characteristics could be inherited. Modern genetic (p. 196) studies, however, have been unable to discover any mechanism whereby characters developed during an individual's lifetime could be passed on to its offspring so that the theory has fallen into disuse.

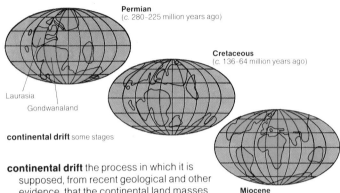

Permian
(*c.* 280-225 million years ago)

Cretaceous
(*c.* 136-64 million years ago)

Laurasia

Gondwanaland

continental drift some stages

Miocene
(*c.* 26-7 million years ago)

continental drift the process in which it is
supposed, from recent geological and other
evidence, that the continental land masses
have not always occupied their present position
on the globe and that, powered by processes
within the Earth itself, they are slowly and
continuously on the move. In this way, new
land masses are created and destroyed, split
apart and joined, over geological time. This is
one of the processes that can lead to the
geographical isolation (p. 214) of a breeding
population (p. 214) and so to speciation (p. 213).

Pangea (*n*) the single landmass or
'supercontinent' which itself was formed during
Devonian times, some 395 to 345 million years
ago, by collision of the two original continents,
known as Gondwanaland and Laurasia. The
present continents evolved from Pangea by the
process of continental drift (↑).

plate tectonics the theory (p. 235) which has
been developed recently to provide a
mechanism for continental drift (↑). It supposes
that the surface layers of the Earth fit together,
rather like a spherical jigsaw puzzle, and that
the individual pieces are on the move in
relation to one another. In this way, pieces may
slide past one another, collide with one piece
being forced beneath the other, or separate
with new crust being formed as they move
apart. It is at the boundaries between the
individual plates that the majority of volcanic
eruptions and earthquakes take place.

analogous structures

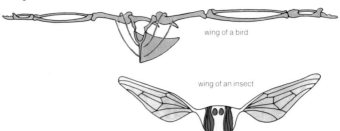

wing of a bird

wing of an insect

homologous structures

flipper of a turtle

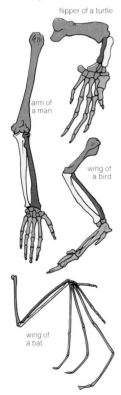

arm of
a man

wing of
a bird

wing of
a bat

analogous (*adj*) of structures or organs which
occur in different species (p. 40) of organisms
and that have similar functions but a different
evolutionary (p. 208) and embryological (p. 166)
origin so that their structure is also different.
For example, the wings of birds and those of
the insects both enable the animals to fly but
their origins and form are quite different.

homologous (*adj*) of structures or organs which
occur in different species (p. 40) of organisms
but which have similar evolutionary (p. 208)
and embryological (p. 166) origins even though
their functions may have been modified. For
example, the limbs of all tetrapod (p. 77)
vertebrates (p. 74) are based on the pattern of
five digits. This suggests evolutionary
relationships between different species.

divergent (*adj*) of evolution (p. 208) in which
homologous (↑) structures have become
adapted to perform different functions. For
example, the flippers of sea mammals (p. 80),
such as seals, are homologous with the limbs
of land-based vertebrates (p. 74) but, are used
in a different way as the seals have become
better adapted to their marine environment
(p. 218).

convergent (*adj*) of evolution (p. 208) in which
analogous (↑) structures have become adapted
to perform the same function. For example, the
eye of a cephalopod (p. 72) performs the same
function as that of a vertebrate (p. 74) but has a
quite different origin and structure.

vestigial (*adj*) of a structure or organ which originally performed a useful function but, through evolution (p.208), has become reduced to a remnant of its former self and no longer functions. For example, the appendix (p. 102) in humans.

primitive (*adj*) of a structure or organism that is at an early stage in evolution (p. 208), or is like an organism at an early stage.

phylogenetic (*adj*) of a classification (p. 40) which is based on the apparent evolutionary (p. 208) relationships between organisms.

palaeontology (*n*) the science or study of ancient life forms through their remains as fossils (↓).

fossil (*n*) any remains or trace of a once-living organism that has been preserved in some way such as in the rocks or in ice.

fossil e.g.
Archaeopteryx

fossil record the continuing record of the origins, development and existence of life on Earth as expressed through the finds of fossils (↑) preserved in the rocks from the origins of the planet to the present day.

geological column a tabular time scale which has been worked out by geologists on the basis of the fossil record (↑) and other evidence, such as radiometric dating, in which the history of the Earth is broken down into eras, periods, and epochs.

variation (*n*) the differences in form and structure which occur naturally among individuals within the same species (p. 40) and which may result from genetic (p. 196) changes, such as mutations (p. 206), or from differences in such factors as nutrition (p. 92) or the density of the population (p. 214).

diversity (*n*) the state of things being different from each other.

isolating mechanisms factors, such as the existence of geographical barriers, behaviour (p.164), or the timing of the breeding season (p.195), which tend to separate groups of individuals into reproductive (p. 173) communities (p. 217).

gene pool the total number and type of genes (p. 196) that exist at any given time within a breeding population (p. 214) that has been separated by various isolating (p. 214) mechanisms. The genes within a given gene pool may then be intermixed randomly by interbreeding within the group.

speciation (*n*) the process by which two or more new species (p. 40) evolve (p. 208) from one original species as breeding groups become separated by isolating (p. 214) mechanisms and develop a range of distinctive characters, as a result of natural selection (p. 208), to the extent that the isolated populations (p. 214) are no longer able to breed with one another.

speciation   time

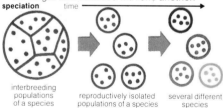

interbreeding
populations
of a species

reproductively isolated
populations of a species

several different
species

differential mortality the basis of natural selection (p. 208) during periods of increasing population (p. 214) when those individuals of the overpopulated community which are best fitted to their environment (p. 218) survive to breed while those that are less well fitted die so that evolution (p. 208) takes place by natural selection.

melanism (*n*) the condition in which such structures as hair, skin and eyes are coloured by the dark-brown pigment (p. 126) melanin. Melanistic skin protects the individual from the harmful effects of prolonged exposure to sunlight. Consequently, humans who have evolved (p. 208) in areas of high sunlight intensity have, by natural selection (p. 208), evolved darker skin colour.

gene frequency the occurrence of one particular gene (p. 196) in a given population (↓) in relation to all its other alleles (p. 197).

Hardy-Weinberg principle a law formulated in 1908 from which the effects of natural selection (p. 208) can be better understood. It suggests that in any population (↓), in which mating takes place at random, the proportion of dominant (p. 197) to recessive (p. 197) genes (p. 196) in the population remains unchanged from one generation to the next. Until the principle was worked out, it was thought, quite reasonably, that the numbers of recessive genes would decline while the dominant genes would increase.

gene flow the process by which genes (p. 196) move within a population (↓) by mating and the exchange of genes.

genetic drift the process by which the genetic (p.196) structure of a small population (↓) of organisms changes by chance rather than by natural selection (p. 208). In a small population the Hardy-Weinberg principle (↑) may not be maintained because the number of pairings will not be random.

isolation (*n*) the process by which two populations (↓) become separated by geographical, ecological (p. 217), behavioural (p. 164), reproductive (p. 173), or genetic (p. 196) factors. After two populations have become genetically or reproductively separated, they will not revert to the same species (p. 40) even if they come together geographically again.

population (*n*) a group of organisms of the same species (p. 40) which occupies a particular space over a given period of time. The actual numbers of individuals within a population may rise and fall as a result in changes of the birth and death rate and such factors as climate, food supply, and disease.

allopatric (*adj*) of two or more populations (↑) of
the same or related species (p. 40) which could
interbreed if they were not geographically
isolated (↑) from one another.

sympatric (*adj*) of two or more related species
(p. 40) which are not geographically isolated (↑)
from one another and which could interbreed
apart from differences in behaviour (p. 164) or
the timing of the breeding season (p. 195) etc.

sympatric
e.g. two species occurring
in the same place

ecological isolation isolation (↑) which occurs
within populations (↑) as a result of the different
ways in which they relate to their environment
(p. 218).

reproductive isolation isolation (↑) which occurs
within populations (↑) as a result of differences
in their breeding behaviour (p. 164) or timing of
their breeding season (p. 195).

allopatric
e.g. two species occurring
in different places

ecological hybridization two related species growing at top and
bottom of a cliff cannot interbreed because of space in between.
On gentle slopes the populations overlap and interbreed

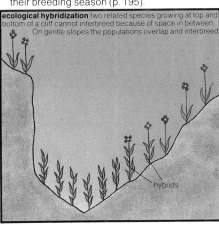

hybrids

genetic isolation isolation (↑) which occurs within
populations (↑) as a result of their genetic
(p. 196) incompatibility so that they are unable to
produce fertile (p. 175) offspring.

artificial selection the process by which humans
make use of the principles of genetics (p. 196)
and evolution (p. 208) to create breeds or
hybrids (p. 216) which would not be expected
to occur as a result of natural selection (p. 208).

inbreeding (*n*) breeding by the mating of closely related individuals, including self-fertilization (p. 175) in plants. It tends to reduce the genetic (p. 196) variability of the population (p. 214) and leads to a greater frequency of expression of recessive (p. 197) characteristics. Humans make use of inbreeding during artificial selection (p. 215) to develop characteristics which are seen as useful.

outbreeding (*n*) breeding by the mating of individuals which are not closely related. The most extreme form of outbreeding is between organisms of different species (p. 40) which leads to the production of non-fertile (p. 175) offspring. Outbreeding normally gives rise to greater genetic (p. 196) variability and vigour and there may be various mechanisms within organisms to encourage it.

hybrid vigour an increase in the vigour of such factors as growth or fertility (p. 175) in the offspring as compared with the parents which results from the cross-breeding of individuals from lines which are genetically (p. 196) different leading to greater heterozygosity (p. 198) and an increased expression of dominant (p. 197) genes.

hybrid (*n*) the offspring of parents from genetically (p. 196) different lines. **hybridization** (*n*).

spontaneous generation the idea, disproved by the French bacteriologist, Louis Pasteur (1822–95) and others, that, in suitable conditions, organisms, especially microorganisms, could be generated from inorganic compounds (p. 15).

special creation a hypothesis (p. 235) which suggests that every form of life that exists or has ever existed was created separately by a deity or other supernatural force. Palaeontological (p. 212) and genetic (p. 196) evidence suggests that this is unlikely and few scientists take the hypothesis seriously today.

steady state a hypothesis (p. 235) which suggests that all organisms were created at some time in the past and have remained unchanged ever since with each generation being identical to its predecessor. Palaeontological (p. 212) evidence suggests that this cannot be the case.

ecology (*n*) the science or study of organisms in relation to one another and the environment (p. 218).

biosphere (*n*) the part of the Earth which includes all of the living organisms on the planet and their environment (p. 218).

biome (*n*) a part of the biosphere (↑) which might be a large, regional community (↓) of interrelated organisms and their environment (p. 218) and would include such habitats (↓) and communities as a tropical rainforest or grassland.

ecosystem (*n*) a self-contained and perhaps small unit or area, such as a woodland, which would include all the living and non-living parts of that unit.

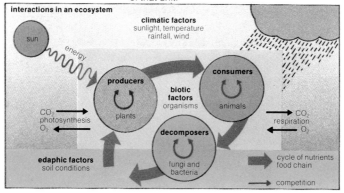

interactions in an ecosystem

climatic factors
sunlight, temperature
rainfall, wind

sun

energy

producers
plants

biotic factors
organisms

consumers
animals

CO_2
photosynthesis
O_2

CO_2
respiration
O_2

decomposers
fungi and bacteria

edaphic factors
soil conditions

cycle of nutrients
food chain

competition

community (*n*) a localized group of a number of populations (p. 214) of different species (p. 40) living and interacting with one another within an ecosystem (↑). Communities can be described as open (niches (p. 219) unstable or 'empty' allowing new species into the community) or closed (niches stable and full).

habitat (*n*) a part of an ecosystem (↑), such as a desert, in which particular organisms live because the environmental (p. 218) conditions within the habitat are essentially uniform even though they may vary with the season or between, say, ground level and the tops of the trees.

microhabitat (*n*) a small area within a habitat (↑) such as the underside of a stone.

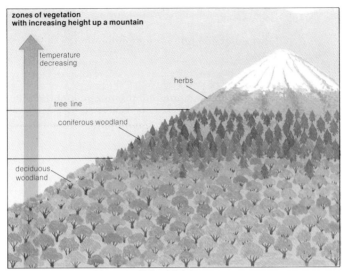

**zones of vegetation
with increasing height up a mountain**

temperature
decreasing

herbs

tree line

coniferous woodland

deciduous
woodland

zone (*n*) a part of a biome (p. 217) which is characterized by one particular group of organisms depending upon the environmental (↓) conditions present in that area.

climate (*n*) the sum total of all the interrelating weather conditions, such as temperature, pressure, rainfall, sunshine, etc., that exists in a particular region throughout the year and averaged over a number of years.

microclimate (*n*) the climate (↑) which occurs in a small region, such as a town or woodland, which differs in some way from the overall climate of the region due to the effects of other factors within the area. For example, the temperature in a large city may be significantly higher than that of its rural surroundings because of the heat that is trapped by the buildings and re-released.

environment (*n*) the sum total of all the external conditions within which an organism lives.

territory (*n*) any area which is occupied and defended by an animal for purposes of breeding, feeding etc.

niche (*n*) the local physical and biological conditions which an organism fills in an ecosystem (p. 217). If, at some stage, more than one species (p. 40) of organism attempt to occupy the same niche, then they compete with one another until one is eliminated. On the other hand, it is possible for different species to occupy the same niche in geographically separated regions or for one species to evolve (p. 208), by natural selection (p. 208), to occupy different niches.

abiotic (*adj*) of the physical environment (↑) to which organisms are subjected, such as temperature, light intensity, availability of water, etc.

aquatic (*adj*) of a watery environment (↑) or a species (p. 40) which lives primarily in water.

freshwater (*adj*) of an aquatic (↑) environment (↑), such as a river, which does not contain salt and is, therefore, not marine (↓). Also, of a species (p. 40) which lives primarily in fresh water.

marine (*adj*) of an aquatic (↑) environment (↑), such as the ocean, which contains salt. Also describes a species (p. 40) which lives primarily in a marine environment.

littoral (*n*) the zone (↑) of a freshwater (↑) environment (↑) between the water's edge and a depth of about six metres or the zone of a marine (↑) environment between the high and low water marks. A littoral species (p. 40) is one which lives primarily in the littoral zone.

amphibious (*adj*) of an organism which is capable of or spends part of its time living in water and part on land.

terrestrial (*adj*) of those organisms that spend most or all of their lives on land.

subterranean (*adj*) of those organisms that spend most or all of their lives underground, in caves, for example.

arboreal (*adj*) of those organisms that spend most or all of their lives living among the branches of trees.

aerial (*adj*) of those organisms or parts of organisms that spend part or all of their lives in the air. The roots of certain trees grow in the air and are referred to as aerial.

climatic factors those aspects of the environment (p. 218), grouped together as the climate (p. 218), including temperature, rainfall, etc., which affect the distribution of organisms.

edaphic factors those aspects of the environment (p. 218) concerned with the soil and including moisture content, pH (p. 15), etc., which affect the distribution of organisms.

biotic (*adj*) of those biological parts of the environment (p. 218) other than the abiotic (p. 219) factors to which organisms are subjected, and include their relationships with other organisms such as competition (↓) for habitat (p. 217) etc.

predation (*n*) the process by which certain animals gain nutrition (p. 92) by killing and feeding upon other animals. A predator is a secondary consumer (p. 223) and predators do not include parasites (p. 110).

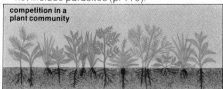

competition in a plant community

leaves compete for light, CO_2, space

roots compete for nutrients and water

competition (*n*) the process in which more than one species (p. 40) or individuals of the same species attempt to make use of the same resources in the environment (p. 218) because there are not enough resources to satisfy the needs of all the organisms. Competition often leads to differential mortality (p. 213).

intraspecific (*adj*) of an action, for example, competition (↑), which takes place between individuals of the same species (p. 40).

interspecific (*adj*) of an action, for example, competition (↑), which takes place between different species (p. 40).

mimicry (*n*) the process in which one organism resembles another and thereby gains some advantage, e.g. a defenceless hoverfly closely resembles the form and colour of a wasp and may, therefore, be avoided by predators (↑).

mimicry

hoverfly

harmless hoverfly mimics unpleasant wasp

wasp

synecology (*n*) the study or science of all communities (p. 217) and ecosystems (p. 217) within an environment (p. 218) and their relationships to one another.

autecology (*n*) the study or science of individuals of one species (p. 40) in relation to one another and to their environment (p. 218).

succession (*n*) a progressive sequence of changes which takes place, after the first colonization (↓) of a particular environment (p. 218), in the organisms which occupy that environment until a stable position is reached where no further changes can take place unless the abiotic (p. 219), edaphic (↑), or climatic factors (↑) themselves are altered. The process takes place rapidly at first and then slows down as stability is approached.

succession

1

a pioneer species colonizes a habitat

2

pioneer plants grow and reproduce

3

growth of plants alters edaphic and biotic factors and more species colonize

5

climax community with many plant species. Conditions no longer suitable for pioneer

colonization (*n*) the arrival and growth to reproductive (p. 173) age of an organism in an area, i.e. the spread of species (p. 40) to places where they have not lived before. **colonize** (*v*), **colony** (*n*).

pioneer (*n*) a plant species (p. 40) that is found in the early stages of succession (↑).

climax (*adj*) of a community (p. 217) which, following succession (↑), has reached stability.

sere (*n*) a succession (↑) of plant communities (p. 217) which themselves affect the environment (p. 218) leading to the next community and resulting ultimately in the climax (↑) community.

soil (*n*) the material which forms a surface covering over large areas of the Earth and in which organisms gain support, protection, and nutrients (p. 92). It results from the weathering and breakdown of rocks into inorganic (p. 15) mineral particles which are then further acted upon by climatic (p. 220) and biotic (p. 220) factors. The composition depends upon the composition of the original rock.

inorganic component the part of the soil which results from the action of weather on the parent rocks, breaking it down into mineral particles of varying size and composition depending upon the composition of the original rock.

organic component the part of the soil which is derived from the existence and activity of the large numbers of living organisms in the soil.

sand (*n*) the inorganic component (↑) in which the particles range in size from 0.02–2.0 millimetres and are angular. A soil with a high sand content tends to be dry, because of the ease with which water drains away, acidic (p. 15), and low in nutrient (p. 92) content.

clay (*n*) the inorganic component (↑) in which the particles are less than 0.02 millimetres in size and are relatively smooth and rounded. A soil with a high clay content tends to be easily waterlogged, can become compacted, and will harden on drying. It is usually rich in nutrient (p. 92) content, however.

humus (*n*) the organic component (↑) of soil which results from the activity and decomposition (↓) of the living organisms within a soil and which is a mixture of fibrous (p. 143) and colloidal materials made up essentially of carbon, nitrogen, phosphorus and sulphur. Humus improves the structure and texture of a soil, helps it to retain water and nutrients (p. 92), and raises the soil's temperature by absorbing more of the sun's energy because of its dark colour.

erosion (*n*) the process by which the products of weathering of a rock or a soil are worn away by the action of wind, running water, or moving ice, etc.

the constituents of loam
soil shaken up with water

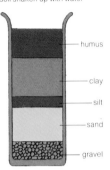

- humus
- clay
- silt
- sand
- gravel

soil profile the series of distinct layers that can be observed in a vertical section through soil from the parent rock, through weathered rock and *subsoil* to *topsoil*.

a generalized soil profile

litter

topsoil, containing humus and minerals

subsoil, containing minerals weathered from rock

parent rock, weathering at the surface

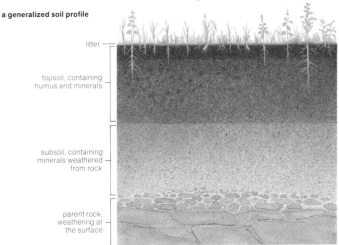

producers (*n.pl.*) the organisms, especially green plants and some bacteria (p. 42), which are able to manufacture nutrients (p. 92) from inorganic (p. 15) sources by such processes as photosynthesis (p. 93).

consumers (*n.pl.*) the heterotrophic (p. 92) organisms which obtain their nourishment by consuming the producers (↑) or other consumers.

decomposers (*n.pl.*) the organisms which obtain their nutrients (p. 92) by feeding upon dead organisms, breaking them down into simpler substances and, in so doing, making other nutrients available for the producers (↑).
decomposition (*n*).

trophic level the particular position which an organism occupies in an ecosystem (p. 217) in respect of the number of steps away from plants at which the organism obtains its food. The producers (↑) are at the lowest trophic level while the predators (p. 220) at the highest trophic levels.

pyramid of available energy at the trophic levels of a food web

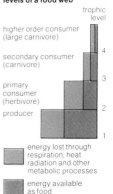

trophic level

higher order consumer (large carnivore)

secondary consumer (carnivore)
4

3
primary consumer (herbivore)

producer
2

1

energy lost through respiration, heat radiation and other metabolic processes

energy available as food

carbon cycle

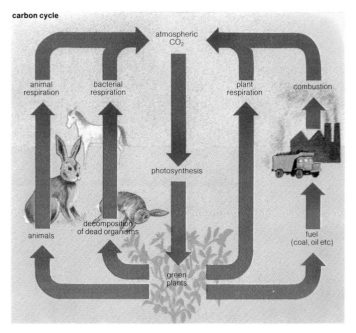

carbon cycle the chain or cycle of events by
which carbon is circulated through the
environment (p. 218) and living organisms.
Plants take in carbon dioxide from the
atmosphere and turn it into carbohydrates
(p. 17), proteins (p. 21) and fats. Some of the
carbon dioxide is returned to the atmosphere
during the plants' respiration (p. 112). The
plants are eaten by herbivores (p. 105) which,
in turn, are eaten by carnivores (p. 105). When
the herbivores and carnivores die, they are fed
upon by saprophytes (p. 92) and decomposers
(p. 223) so that carbon is returned to the soil
or to the atmosphere as a product of respiration
of bacteria (p. 42) and fungi (p. 46).

oxygen cycle the chain or cycle of events by
which oxygen is circulated through the
environment (p. 218) and living organisms.

nitrogen cycle the chain or cycle of events by which nitrogen is circulated through the environment (p. 218) and living organisms. Some bacteria (p. 42) and algae (p. 44) can make use of nitrogen directly, and lightning, acting upon atmospheric nitrogen and oxygen, causes it to combine into nitrous and nitric oxide which dissolve in falling rain to enter the soil and form nitrates and nitrites. Most plants make use of nitrogen as nitrates and use them in the manufacture of proteins (p. 21). The plants are fed upon by herbivores (p. 105) which, in turn, are eaten by carnivores (p. 105) which also make use of the nitrogen in the manufacture of animal proteins. When animals and plants die, the nitrogen is returned to the soil by nitrifying bacteria as nitrites, ammonia, and ammonium compounds.

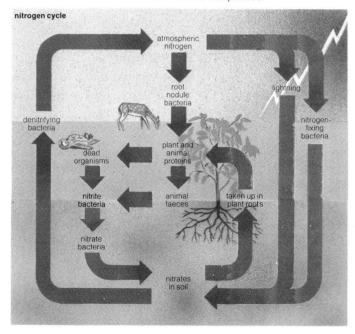

nitrogen cycle

water cycle the chain or cycle of events by which water, essential for life, is circulated through the environment (p. 218) and living organisms.

food chain the sequence of organisms from producers (p. 223) to consumers (p. 223) which feed at different trophic levels (p. 223). A simple food chain: grass grows; a cow eats the grass; a human eats the cow or drinks its milk.

food web an interconnected group of food chains (↑). There are few systems as simple as a food chain and many chains may interlink to form a complex web.

a food chain

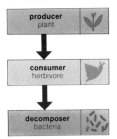

a food web

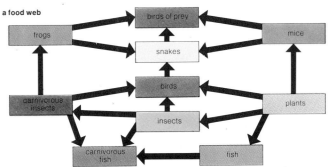

biomass (n) the total mass or volume of all the living organisms within a particular area, community (p. 217), or the Earth itself.

pyramid of biomass a diagrammatic representation, which forms the shape of a gently sloping pyramid, to show the biomass (↑) at every trophic level (p. 223).

standing crop the total amount of nutritional (p. 92) living material in the biomass (↑) of a given area at a particular time.

diurnal rhythm the rhythmic sequence of metabolic (p. 26) events, such as the motion of leaves in plants, that take place over a roughly twenty-four hour pattern, and which can be shown to occur in all living organisms even if they are isolated from their normal external environment (p. 218).

circadian rhythm = diurnal rhythm (↑).

annual rhythm the rhythmic sequence of
metabolic (p. 26) events, such as germination
(p. 168), flowering and fruiting in plants, that
take place over a roughly yearly pattern even if
they are isolated from their normal external
environment (p. 218).

zooplankton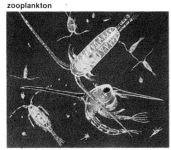

phytoplankton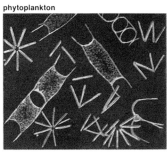

plankton (*n*) any of the various, usually tiny or
microscopic (p. 9), organisms that float freely in
an aquatic environment (p. 218) that have no
visible means of locomotion (p. 143) and
depend on the currents in the water for
distribution. They are not attached to any other
organism or substrate.

phytoplankton (*n*) the plant plankton (↑),
especially the diatoms, which are an important
source of food for other organisms such as
many species (p. 40) of whales.

zooplankton (*n*) the animal plankton (↑) including
the larvae (p. 165) of many species (p. 40) of fish.

pelagic (*adj*) of the upper waters of an aquatic,
especially marine environment (p. 218), as
opposed to the bed of the ocean or lake, and
the organisms which inhabit them.

benthic (*adj*) of the bed of an aquatic, especially
marine environment (p. 218), and the organisms
which live on or in it.

association (*n*) any relationship which exists
between organisms to the benefit of one or all
of them. In plants, a climax (p. 221) community
(p. 217) dominated by one or a small number of
species (p. 40) and named after them. *See also*
parasitism (p. 110).

symbiosis (*n*) an association (p. 227) between two or more species (p. 40) of organisms to their mutual benefit, such as the association of the mycorrhiza (p. 49) of certain fungi (p. 46) with the roots of trees whereby the tree provides nutrients (p. 92) for the fungus which helps the tree to take up water and supplies nitrates to the roots. **symbiotic** (*adj*).

ectotrophic mycorrhiza

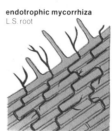
endotrophic mycorrhiza
L.S. root

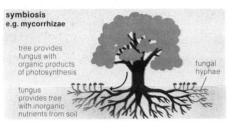
symbiosis
e.g. mycorrhizae

tree provides
fungus with
organic products
of photosynthesis

fungal
hyphae

fungus
provides tree
with inorganic
nutrients from soil

commensalism (*n*) an association (p. 227) in which one species (p. 40) of organism, the commensal, benefits while the other species is neither harmed nor gains benefit. The bacteria (p. 42) in the gut (p. 98) of mammals (p. 80) are commensals.

mutualism (*n*) an association (p. 227) between two or more species (p. 40) in which both benefit. In some cases of mutualism, neither species may be able to survive without the other while, in others, both species may be able to survive independently. It is a form of symbiosis (↑). For example, a species of sea anemone lives on the back of the hermit crab and benefits from being transported to new feeding sites where it feeds on debris from the crab's meals while the crab is protected from predation (p. 220) by the stinging tentacles of the anemone.

epiphyte (*n*) any plant, such as some ferns or lichens (p. 49), which grows on another plant, in a commensal (↑) association (p. 227), using it only for support and not involved in any parasitism (p. 110).

epizoite (*n*) any animal, such as the remora fish which is attached to a shark by a powerful sucker, which lives permanently on another animal, using it for transport etc. and not involved in any parasitism (p. 110).

farming (*n*) the process in which humans exploit naturally occurring plants and animals to provide food for their own needs either by deliberately cultivating wild species (p. 40) or by developing new types of organisms and then sowing, planting, tending and protecting them.

fishery (*n*) the process in which humans catch fish or other aquatic animals for food, and exploit the natural processes of population (p. 214) control to increase the size of the catch.

maximum sustainable yield the maximum size of catch of, for example, fish that can be obtained and sustained over years from a given area of water which is fished in such a way that the stocks are larger than they would be if they were unfished. The adult fish are removed from the water for food so that the young do not have to compete with the adults to the same degree for food and the biomass (p. 226) of the water is increased by their survival.

biological control
e.g. ladybirds are introduced to control aphids (pests)

aphid (pest)

agriculture (*n*) all of the processes associated with the growing of food in a systematic way, including cultivation of land, tending of stock, development of new types, and destruction of competing (p. 220) species (p. 40), so that the yield from a given area can be increased to cope with the increasing demands of a growing human population (p. 214).

pest (*n*) any species (p. 40) of animal or plant which, in the light of modern methods of agriculture (↑) where vast tracts of land are given over to one species, is not subject to the controls of a natural ecosystem (p. 217) and may increase rapidly in numbers to destroy the crop.

weed (*n*) any species (p. 40) of plant which may be able to grow in an area which has been given over to the cultivation of food plants and which will compete with those food plants for space, light, water and nutrients (p. 92).

ladybird

biological control a method of reducing the numbers of weeds (↑) or pests (↑) by introducing a natural predator (p. 220) of the pest species (p. 40) If the predator is also able to feed on species which are not regarded as pests, then its numbers will not be reduced when the numbers of pests have fallen. This attempts to maintain a natural equilibrium between the pest and the predator.

pesticide (*n*) any agent, usually chemical, which is used to control and destroy pests (p. 229).

herbicide (*n*) any agent, usually chemical, which is used to destroy or control weeds (p. 229).

water purification all of the processes, including storage, straining, filtering and sterilizing which are used by the water authorities to maintain drinking water fit for human consumption. Since drinking water is drawn from rivers, lakes and underground wells, it is also important to ensure that pollutants (↓) from industry or agriculture (p. 229) do not enter the supplies to unacceptable levels.

sewage treatment all of the processes, including the removal of sludge by sedimentation, screening to remove large particles of waste, biological oxidation (p. 32), removal of grit, filtering etc, to ensure that the effluent, which would otherwise contain human waste etc, can be returned to the water cycle (p. 226) without the risk of spreading diseases etc.

conservation (*n*) the use of the natural resources in such a way that they are not despoiled. It is usually taken to include the act of study, management and protection of ecosystems (p. 217), habitats (p. 217) or species (p. 40) of organisms in order to maintain the natural balance of wildlife and its environment (p. 218).

endangered species any species (p. 40) of animal cr plant which, by changes in the natural environment (p. 218) or by human intervention, are threatened with death and extinction.

over-exploitation the use of natural resources in such a way that natural ecosystems (p. 217) may be irreversibly disturbed, habitats (p. 217) destroyed, or organisms threatened with extinction.

pollution (*n*) the act of introducing into the natural environment (p. 218) any substance or agent which may harm that environment and which is added more quickly than the environment is able to render it safe. **pollutant** (*n*), **pollute** (*v*).

water pollution the pollution (↑) of marine and freshwater habitats (p. 217) by the unthinking introduction of human, agricultural (p. 229), and industrial waste into rivers, lakes, and oceans.

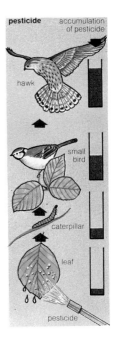

pesticide — accumulation of pesticide

hawk

small bird

caterpillar

leaf

pesticide

principle natural resources exploited by man

oxygen demand the condition which exists in aquatic environments (p. 218) into which pollutants (↑) have been introduced which promote the growth of aerobic (p. 32) bacteria (p. 42) causing a depletion of the levels of oxygen in the water. Thus, the natural plant life of the environment is reduced and, with it, the animal life that depends upon the plants.

eutrophication (*n*) the situation which occurs when an excess of nutrients (p. 92) is introduced into a freshwater habitat (p. 217) causing a dramatic growth in certain kinds of algae (p. 44). When the nutrients have been used up, the algae die, and the bacterial (p. 42) decomposers (p. 223) which feed on the dead algae use up the oxygen in the water giving rise to an oxygen demand (↑).

algal bloom the dramatically increased population (p. 214) of algae which occurs in an aquatic environment (p. 218) which occurs as a result of eutrophication (p. 231).

air pollution pollution (p. 230) of the atmosphere which results from burning fossil fuels, such as coal and oil, with the introduction into the air of organic (p. 15) and inorganic (p. 15) compounds, such as carbon dioxide, carbon monoxide, sulphur dioxide, etc.

smog (n) fog which has been polluted (p. 230). A mixture of smoke and fog.

marine pollution the pollution (p. 230) of the marine environment (p. 218) primarily by crude oil as a result of the illegal washing of tanks at sea or by accidental loss. The damage to seabird populations (p. 214) is great and well known but there is also poisoning of marine plankton (p. 227) which thereby affects the whole marine food web (p. 226).

radioactive pollution the pollution (p. 230) of the environment (p. 218) by accidental leakage from sites of nuclear energy production or from the dumping of nuclear waste products. The radioactive materials which find their way into the environment can lead to chromosome (p. 13) damage and mutations (p. 206).

terrestrial pollution the pollution (p. 230) of the land environments (p. 218) by the dumping of waste materials from mining industries, for example, or by pesticides (p. 230).

birth control the attempt by humans artificially to limit the rapid growth which has taken place in the world human population (p. 214) which may otherwise place possibly disastrous strains on food supplies and other non-replaceable resources. It involves methods of preventing conception with the use of contraceptives such as the Pill, vasectomy, etc.

hygiene (n) the science which deals with the preservation of human health by such means as improvements of sanitation to prevent the spread of disease. It is thought that hygiene improvements are among the most important factors in the increase in human life expectancy.

air pollution smog

pollution of water

disease (*n*) any disorder or illness of the body or organ.

infectious (*adj*) of a disease caused by viruses (p. 43) or other parasitic (p. 92) organisms, such as certain bacteria (p. 42), which can be passed from one individual to another.

contagious (*adj*) of a disease which can be passed from one individual to another by contact, which may be direct touching of the individuals or through objects which have been contaminated by the diseased individual and then handled by another individual.

antiseptic (*adj*) of any agent which destroys the microorganisms which invade the body leading to disease.

aseptic (*adj*) of conditions in which disease-causing microorganisms are not present.

antibiotic (*n*) any substance, produced by a living organism, for example, the fungus (p. 46) *Penicillium*, which is poisonous to other living organisms. Antibiotic substances are used in medicine to destroy disease-causing microorganisms.

antibody (*n*) a protein (p. 21) which is produced by an organism following the invasion of the body fluids by a substance which is not normally present and which may be harmful. The antibody combines with the invading substance thereby removing it from the body.

immunity (*n*) the state in which organisms are protected from the invasion of disease which mainly involves the production of antibodies (↑).

active immunity immunity (↑) in which the body's defensive mechanisms are stimulated by the invasion of foreign microorganisms to produce antibodies (↑).

passive immunity immunity (↑) in which the body's own defensive mechanisms are not stimulated by the invasion of foreign micro-organisms but in which antibodies (↑) have been transferred to it from another animal in which active immunity (↑) has been stimulated.

inherited immunity passive immunity (↑) in which the resistance to certain diseases is inherited genetically (p. 196) from the parents.

the production
of antibodies

harmful substances
invade animal

defence mechanisms of
animal produce antibodies

antibodies
combine with
harmful substances

combinations of antibody
and now harmless substances
removed from the body

acquired immunity active immunity (p. 233) by exposure to an infectious (p. 233) disease which is too restricted to cause the symptoms of the disease or passive immunity (p. 233) by the transfer of antibodies (p. 233) from the mother to the offspring across the placenta (p. 192).

vaccination (*n*) the injection into the body of an animal modified forms of the microorganisms which will cause a particular disease so that the body produces antibodies (p. 233) that will resist any possible invasion of the disease itself. The animal gains acquired immunity (↑).

vaccine (*n*) any substance containing antigens (↓) which is injected into an animal's body to produce antibodies (p. 233) and give the animal acquired immunity (↑) to specific diseases.

antigen (*n*) any substance, produced by a micro-organism, which will stimulate the production of antibodies (p. 233).

epidemic (*adj*) of a disease which is not normally present in a population (p. 214) and which, therefore, will spread rapidly from individual to individual and infect (p. 233) a large number of the population because there is no natural immunity (p. 233) to the infection.

endemic (*adj*) of a disease which occurs naturally in particular, geographically restricted populations (p. 214).

pandemic (*adj*) of a disease which occurs throughout the population (p. 214) of a whole continent or even the world.

allergy (*n*) the condition in which certain individuals may be particularly sensitive to substances which are quite harmless to other individuals. For example, asthmatic attacks may be stimulated by breathing dust or pollen (p. 181). Allergic reactions may include inflammation or swelling.

symptom (*n*) a sign or condition of the presence of, e.g. a disease.

contract (*v*) (1) *of diseases* to get or to catch. (2) to become smaller or shorter e.g. muscles (p. 143) contract. **contraction** (*n*), **contractile** (*adj*).

acquired immunity

animal with disease-carrying microorganisms

a few microorganisms injected into another animal

antigens produced by the microorganisms cause animal to produce antibodies

animal has gained an acquired immunity to the disease through vaccination

scientific method a means of gaining knowledge of the environment (p. 218) by observation (↓) which leads to the development of a hypothesis (↓). From the hypothesis, predictions (↓) are made which are tested by experiments (↓) that include controls (↓).

observation (*n*) (1) a natural event or phenomenon which is viewed or learned; (2) that which is viewed or learned.

hypothesis (*n*) an idea which has been put forward to explain the occurrence of a natural event or events noted by observation (↑). **hypotheses** (*pl.*).

prediction (*n*) the process of foretelling likely events or phenomena in a given system from those already noted by observation (↑). Predictions follow from the hypothesis.(↑).

experiment (*n*) a means of examining a hypothesis (↑) by testing a prediction (↑) made on the basis of the hypothesis.

control (*n*) an experiment (↑) performed at the same time as the main experiment which differs from it in one factor only. Controls are a means of testing those factors which affect a phenomenon.

theory (*n*) an idea or set of ideas resulting from the scientific method (↑) used as principles (↓) to explain natural phenomena which have been noted by observation (↑).

phenomenon (*n*) any observable fact that can be described scientifically. **phenomena** (*pl.*).

principle (*n*) a general truth or law at the centre of other laws.

adaptation (*n*) a change in structure, function etc, which fits a new use. A particular adaptation may make an organism better fitted to survive (p. 209) in its environment (p. 218). **adapt** (*v*).

structure (*n*) the way in which all the parts of an object, or organism or part of an organism are arranged. The structure of anything is closely related to the function it performs.

function (*n*) the normal action of an object or part of an organism, for example, the function of the ear (p. 157) is to hear (p. 159).

adjacent (*adj*) near by, next to or close to.

amorphous (*adj*) without shape or form, e.g. cells which have not been differentiated.

anterior (*adj*) at, near or towards the front (or head) end of an animal, usually the end directed forward when the animal is moving (in humans the anterior is the ventral (p. 75) part).

articulation (*n*) the movable or non-movable connection or joint between two objects.

axis (*n*) a real or imaginary straight line about which an object rotates e.g. the axis of symmetry (p. 60).

cavity (*n*) a hole or space e.g. the buccal cavity (p. 99).

comatose (*adj*) inactive and in a deep sleep as, for example, in a hibernating (p. 132) animal.

comparable (*adj*) of two or more objects, of similar quality. **compare** (*v*).

concentration (*n*) the strength or quantity of a substance in, for example, a solution (p. 118).

constituent (*n*) a part of the whole. **constituent** (*adj*).

constrict (*v*) to make thinner, for example, a narrowing of the blood vessels (p. 127). **constriction** (*n*).

dilate (*v*) to make wider, for example, in blood vessels (p. 127). **dilation** (*n*).

convoluted (*adj*) rolled or twisted into a spiral or coil, for example, the convoluted tubules in the kidney (p. 136). **convolute** (*v*).

co-ordinate (*v*) to cause two or more things, e.g. limbs, to work together for the same purpose. **co-ordination** (*n*).

crystallize (*v*) to form crystals (regular shapes).

deficiency (*n*) a shortage or lack of something. For example, vitamin deficiency (*see* p. 238).

development (*n*) a stage in growth which includes changes in structures and the appearance of new organs and tissues (p. 83).

duct (*n*) a tube formed of cells.

equilibrium (*n*) the state in which an object is steady or stable because the forces acting upon it are equal.

essential (*adj*) very necessary.

external (*adj*) of the outside.

internal (*adj*) of the inside.

extract (*v*) to remove or draw out one substance from a particular material.

filter (*n*) an instrument used to take solids and other substances out of liquids. **filtration** (*n*).

flex (*v*) *of a joint* to bend, *of a muscle* (p. 143) to contract.

gradient (*n*) the increase or decrease in a substance over a distance.

increase (*v*) to become or to make greater in some way, for example, in size, value, concentration etc. **increase** (*n*).

decrease (*v*) to become or to make less or fewer in some way, for example, in size, value, concentration etc. **decrease** (*n*).

insulation (*n*) any material used to prevent the passage of heat (or electricity), for example, hair insulates the bodies of mammals (p. 80) and feathers (p. 147) the bodies of birds.

intermediate (*adj*) of an object in the middle, e.g. an intermediate stage in metabolism (p. 26).

lubricate (*v*) to make smooth or slippery in order to make the movement of parts of a machine or organism easier. **lubrication** (*n*).

offspring (*n*) = progeny (p. 200).

parallel (*adj*) of lines or planes which run in the same direction and never meet.

permeable (*adj*) of, for example, a membrane (p. 14) which allows a substance to pass through. *See also* semipermeable membrane (p. 118).

posterior (*adj*) at, near or towards the back or hind end of an animal, usually the end directed backwards when the animal is moving.

product (*n*) a substance that is produced.

byproduct (*n*) a substance that is produced in the course of producing another substance.

protuberance (*n*) a part or thing which swells or sticks out, for example, a pseudopodium (p. 44) of *Amoeba* (p. 44).

sedentary (*adj*) of an animal that remains attached to a surface and does not carry out locomotion (p. 143), for example, a coral polyp (p. 61).

synthesize (*v*) to make a substance from its parts.

tensile (*adj*) of a material which is able to be stretched.

transparent (*adj*) of a material that lets light pass through and through which objects can be clearly seen.

viscous (*adj*) of a fluid that will not flow i.e. is rather solid.

Vitamins

NAME	LETTER	MAIN SOURCES	FUNCTION	EFFECTS OF DEFICIENCY	FAT (F) OR WATER (W) SOLUBLE
retinol	A	liver, milk, vegetables containing yellow and orange pigments e.g. carrots	light perception, healthy growth, resistance to disease	night blindness, poor growth, infection, drying and degeneration of the cornea	F
calciferol	D	fish liver, eggs, cheese, action of sunlight on the skin	absorption of calcium and phosphorus and their incorporation into bone	bone disorders e.g. rickets	F
tocopherol	E	many plants, such as wheatgerm and green vegetables	cell respiration, conservation of other vitamins	in humans, no proved effect, may cause sterility, muscular dystrophy in rats	F
phylloquinone	K	green vegetables, egg yolk, liver	synthesis of blood clotting agents	haemorrhage, prolonged blood clotting times	F
thiamin	B_1	most meats and vegetables, especially cereals and yeast	coenzyme in energy metabolism	beri-beri, loss of apetite and weakness	W

riboflavine	B_2	milk, eggs, fish, green vegetables	coenzyme in energy metabolism	ulceration of the mouth, eyes and skin	W
niacin	B complex (B_2)	fish, meat, green vegetables, wheatgerm	coenzyme in energy metabolism	pellagra: skin infections, weakness, mental illness	W
pantothenic acid	B_5	most foods, especially yeast, eggs, cereals	coenzyme in energy metabolism	headache, tiredness, poor muscle co-ordination	W
pyridoxine	B_6	most foods, especially meat, cabbage, potatoes	release of energy, formation of amino acids	nausea, diarrhoea, weight loss	W
biotin	B complex (H)	most foods, especially milk, yeast, liver, egg yolk	coenzyme in energy metabolism	dermatitis	W
folic acid	B_c	green vegetables, liver, kidneys	similar to vitamin B_{12}	a form of anaemia	W
cobalamin	B_{12}	meats e.g. liver, heart, herrings, yeast, some green plants	maturing red blood cells, growth, metabolism	a form of anaemia	W
ascorbic acid	C	citrus fruits, green vegetables	collagen formation	scurvy: tooth loss, weakness susceptibility to disease, weight loss	W

Nutrients

carbon dioxide a colourless, odourless gas at normal temperature and pressure with the chemical formula CO_2. It is denser than oxygen and occurs in the atmosphere at lower levels. It is absorbed by plants and is used to make complex organic compounds especially by photosynthesis. It is a waste product of respiration.

oxygen a colourless, odourless gas at normal temperature and pressure with the chemical formula O_2. It is a vital element of the inorganic and organic compounds, such as carbohydrates, proteins and fats, which make up all living organisms. It is taken in by plants as gaseous oxygen in the dark and as carbon dioxide and water and released as a gas from photosynthesis. It is essential for respiration in aerobic organisms.

water a colourless, tasteless liquid, at normal temperatures and pressures, with the chemical formula H_2O. Most nutrients are soluble in water. Water takes part in many of the chemical reactions involved in nutrition and is also an essential fluid in the transport of materials throughout the body of an organism. It is a waste product of respiration and is essential in photosynthesis.

PLANT NUTRIENTS

macronutrients

potassium a macronutrient which is absorbed by plants in the form of potassium salts and which is required as a component of enzymes and amino acids. Potassium deficiency will eventually lead to the plant's death and is indicated by yellow edges to the leaves.

calcium a macronutrient which is absorbed by plants in the form of calcium salts and which is required in cell walls. Calcium deficiency will cause a plant to have stunted roots and shoots because the growing points die.

nitrogen a macronutrient present in the atmosphere as a colourless, odourless gas at normal temperatures and pressures but absorbed by plants in the form of nitrates. It is an essential part of proteins and amino acids etc. Nitrogen deficiency causes the plant to show stunted growth with yellowing of the leaves.

phosphorus a macronutrient which is absorbed by plants as H_2PO_4 and is found in proteins, ATP and nucleic acids. Phosphorus deficiency causes the plant to show stunted growth with dull dark green leaves.

magnesium a macronutrient which is absorbed by plants in the form of magnesium salts and is found in chlorophyll. Magnesium deficiency causes yellowing of the leaves.

sulphur a macronutrient which is absorbed by plants as sulphates and is found in certain proteins. Sulphur deficiency causes roots to develop poorly as well as yellowing of the leaves.

iron a macronutrient which is absorbed by plants as iron salts and is found in cytochromes. Iron deficiency causes yellowing of the leaves.

micronutrients

boron a micronutrient absorbed by plants in the form of borates. It is important after pollination in the stimulation of germination of the pollen grains as well as in the absorption of calcium through the roots. Boron deficiency results in certain diseases of plants, such as internal cork in apples.

zinc a micronutrient which is absorbed by plants in the form of zinc salts. It is important in the activation of certain enzymes and in the production of leaves. Zinc deficiency results in the abnormal growth of leaves.

copper a micronutrient which is absorbed by plants in the form of copper salts. It is required by some enzymes. Copper deficiency results in the growth of the plant showing certain kinds of abnormality.

molybdenum a micronutrient absorbed by plants in the form of molybdenum salts. It is important in the function of certain enzymes for the reduction of nitrogen. Molybdenum deficiency results in the overall growth of the plants being reduced.

chlorine a micronutrient which is absorbed by plants in the form of chlorides. It is important in osmosis etc. although it cannot easily be shown to have effects if there is a deficiency.

manganese a micronutrient which is absorbed by plants in the form of manganese salts. It is an important activator of certain enzymes. Manganese deficiency results in the yellowing of the leaves as well as grey mottling.

ANIMAL NUTRIENTS

minerals

calcium a mineral, present in milk products, fish, hard water and in bread, which is required for healthy bones and teeth to aid in the clotting of blood and in muscles. The average adult human requires 1.1 grams per day and the total body content is about 1000 grams.

phosphorus a mineral, present in most foods but especially cheese and yeast extract, which is required for healthy bones and teeth, and takes part in the DNA, RNA and ATP metabolism. The average human adult requires 1.4 grams per day and the total body content is about 780 grams.

sulphur a mineral, present in foods containing proteins, such as peas, beans and milk products. It is required as a constituent of certain proteins, such as keratin and vitamins, such as thiamine. The average human adult requires 0.85 grams per day and the total body content is about 140 grams.

potassium a mineral present in a variety of foods, such as potatoes, mushrooms, meats and cauliflower, which is required for nerve transmission acid-base balance. The human requires 3.3 grams per day and the total body content is about 140 grams.

sodium a mineral present in a variety of 'salty' foods but especially table salt (sodium chloride), cheese and bacon, and which is required for nerve transmission and acid-base balance. The average human requires about 4.4 grams per day and the total body content is about 100 grams.

chlorine as chloride ions, a mineral found with sodium in table salt and in meats, which is required for acid-base balance and for osmoregulation. The average human adult requires 5.2 grams per day and the total body content is about 95 grams.

magnesium a mineral present in most foods, but especially cheese and green vegetables, which is required to activate enzymes in metabolism. The average human adult requires about 0.34 grams per day and the total body content is about 19 grams.

iron a mineral present in liver, eggs, beef and some drinking water, which is an essential constituent of haemoglobin and catalase. The average human adult requires 16 milligrams per day and the total body content is about 4.2 grams.

fluorine as fluoride, a mineral found in sea water and sea foods and sometimes added to drinking water. It is a constituent of bones and teeth and prevents tooth decay. The average human requires 1.8 milligrams per day and the total body content is about 2.6 grams.

zinc a mineral found in most foods, but especially meat and beans, which is required as a constituent of many enzymes. It is also thought to promote healing. The average human adult requires 13 milligrams per day and the total body content is about 2.3 grams.

copper a mineral found in most foods, but especially in liver, peas and beans, which is required for the formation of haemoglobin and certain enzymes. The average human adult requires 3.5 milligrams per day and the total body content is about 0.07 grams.

iodine a mineral found in sea foods and some drinking water and vegetables, which is required as a constituent of thyroxine. The average adult human requires 0.2 milligrams per day and the total body content is only about 0.01 grams.

manganese a mineral found in most foods, but especially tea and cereals, which is required in bones and to activate certain enzymes in amino acid metabolism. The average adult human requires 3.7 milligrams per day and the total body content is only 0.01 grams.

chromium a mineral found in meat and cereals.

cobalt a mineral found in most foods, but especially meat and yeast products, which is an essential constituent of vitamin B_{12}. The average adult human requires 0.3 milligrams per day and the total body content is as little as 0.001 grams.

International System of Units (SI)

PREFIXES

PREFIX	FACTOR	SIGN	PREFIX	FACTOR	SIGN
milli-	$\times 10^{-3}$	m	kilo-	$\times 10^{3}$	k
micro-	$\times 10^{-6}$	μ	mega-	$\times 10^{6}$	M
nano-	$\times 10^{-9}$	n	giga-	$\times 10^{9}$	G
pico-	$\times 10^{-12}$	p	tera-	$\times 10^{12}$	T

BASIC UNITS

UNIT	SYMBOL	MEASUREMENT
metre	m	length
kilogram	kg	mass
second	s	time
ampere	A	electric current
kelvin	K	temperature
mole	mol	amount of substance

DERIVED UNITS

UNIT	SYMBOL	MEASUREMENT
newton	N	force
joule	J	energy, work
hertz	Hz	frequency
pascal	Pa	pressure
coulomb	C	quantity of electric charge
volt	V	electrical potential
ohm	Ω	electrical resistance

Index

Glycolysis

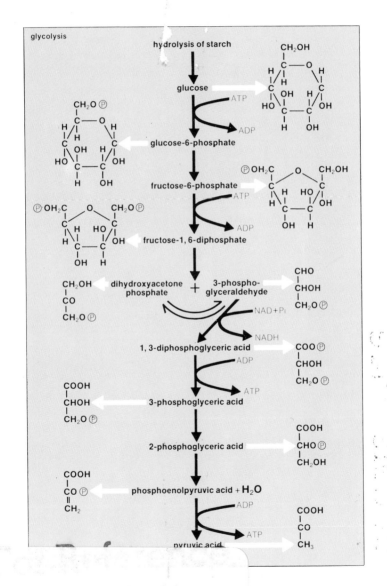

glycolysis

hydrolysis of starch

glucose

ATP

ADP

glucose-6-phosphate

$CH_2O \circledP$

fructose-6-phosphate

ATP

ADP

fructose-1, 6-diphosphate

dihydroxyacetone phosphate + 3-phospho-glyceraldehyde

CH_2OH
CO
$CH_2O \circledP$

CHO
$CHOH$
$CH_2O \circledP$

NAD+Pi

NADH

1, 3-diphosphoglyceric acid

$COO \circledP$
$CHOH$
$CH_2O \circledP$

ADP

ATP

3-phosphoglyceric acid

$COOH$
$CHOH$
$CH_2O \circledP$

2-phosphoglyceric acid

$COOH$
$CHO \circledP$
CH_2OH

phosphoenolpyruvic acid + H_2O

$COOH$
$CO \circledP$
$\parallel$
CH_2

ADP

ATP

pyruvic acid

$COOH$
CO
CH_3